W9-DEX-749

her Titles of Interest

PERGAMON INTERNATI
of Science, Technology, Engineeri
The 1000-volume original paperback l
industrial training and the enjo
Publisher: Robert Maxw

O

B(
Cl
H
H
M
T

STELLAR EVOL

SECOND EDITION

THE PERGAMON TEXTBO
INSPECTION COPY SERVIC

An inspection copy of any book published in the Pergamon Internatio
be sent to academic staff without obligation for their consideration f
recommendation. Copies may be retained for a period of 60 days from
if not suitable. When a particular title is adopted or recommended for
and the recommendation results in a sale of 12 or more copies, the
be retained with our compliments. The Publishers will be pleased to re
revised editions and new titles to be published in this important Intern;

STELLAR EVOLUTION

BY

A. J. MEADOWS
*Department of Astronomy and History of Science,
University of Leicester*

SECOND EDITION

PERGAMON PRESS
OXFORD · NEW YORK · TORONTO · SYDNEY
PARIS · FRANKFURT ·

U. K.	Pergamon Press Ltd., Headington Hill Hall, Oxford OX3 0BW, England
U. S. A.	Pergamon Press Inc., Maxwell House, Fairview Park, Elmsford, New York 10523, U.S.A.
C A N A D A	Pergamon of Canada Ltd., 75 The East Mall, Toronto, Ontario, Canada
A U S T R A L I A	Pergamon Press (Aust.) Pty. Ltd., 19a Boundary Street, Rushcutters Bay, N.S.W. 2011, Australia
F R A N C E	Pergamon Press SARL, 24 rue des Ecoles, 75240 Paris, Cedex 05, France
FEDERAL REPUBLIC OF GERMANY	Pergamon Press GmbH, 6242 Kronberg-Taunus, Pferdstrasse 1, Federal Republic of Germany

First edition 1967

Second edition 1978

British Library Cataloguing in Publication Data

Meadows, Arthur Jack
Stellar evolution. — 2nd ed. — (Pergamon international library).
1. Stars — Evolution
I. Title
523.8   QB806   77-30698

ISBN 0-08-021668-4 Hard cover
ISBN 0-08-021669-2 Flexi cover

Printed in Great Britain by Biddles Ltd., Guildford, Surrey

Contents

Preface to the Second Edition

STARS are the basic building blocks of the Universe. The way in which they change with time determines the entire appearance of the heavens. One of the great advances in science since the Second World War has been the development of a detailed understanding of stellar evolution. This book is intended to describe our present picture of the birth, life and death of stars in a way that is comprehensible to the non-specialist.

CHAPTER 1

The Characteristics of Stars

Classification

The study of stellar evolution has much in common with the study of evolution in biology. Indeed, research work in these two quite different fields started at about the same time — in the latter half of the last century. The approach is the same in both cases. First of all the characteristics of as large a number of plants (or animals, or stars) as possible are studied. Then certain of these characteristics are selected and a system of classification is built up using them as a basis. The most important step, of course, lies in selecting the right characteristics, so that the resulting classification has some physical significance. The next stage is only slightly less important. This consists of using these characteristics to provide a systematic order. For example, you might decide that a significant characteristic of plants is whether they shed their leaves, or retain them throughout the winter. You might then choose this as the first way of dividing plants into two groups. Each group could then be examined again and further subdivided on the basis of other significant characteristics, until the whole supply of significant characteristics has been used. The series of groups with which you end then form the core of your system of classification.

So far there is no question of evolution. In fact, the process of classification presupposes that the objects classified do not change their properties. However, it has been found both in biology and in astronomy that there are always a few objects which do not fit satisfactorily into any grouping. They exist untidily somewhere in between. The study of these odd objects — the so-called 'missing links' — is an important part of astronomy, as of biology. They are indications that a static picture of the Universe is wrong; we must seek instead for gradual changes in characteristics.

Although the concept of evolution in biology and astronomy has many similarities, there is also an important distinction. In biology, evolutionary change occurs in the group: the way in which an individual changes from his birth to his death is not considered as evolution. In astronomy, on the other hand, both the way in which specific stars change, and the way in which groups of stars change, are called evolution. In fact, in this book the main theme will be the individual star, and how it changes with time. By a star we will mean from now on a very large, very hot sphere of material suspended in space.

The first necessity is to decide which of the characteristics of stars are fundamental and which are of secondary importance. We will spend the rest of this chapter looking at the most obvious stellar characteristics and deciding which are likely to be important for stellar evolution.

Brightness and Distance

Stars differ most obviously in their brightness. A brief glance at the night sky is sufficient to show this. Unfortunately, the brightness of a star cannot be measured directly, for it depends not only on the intrinsic properties of the star, but also on its distance away from the observer. The headlamps of a car may seem quite faint when it is a mile away, but they can be dazzlingly bright when it is within a hundred yards So, if we want to find out how bright the stars really are, we must first find out how far away they are. Distance measurements, however, are some of the most difficult to make in astronomy. Moreover, a given method is usually only applicable to a restricted number of stars. More than once in the history of astronomy our beliefs concerning the Universe in which we live have been led astray by inaccurate measurements of distance.

There are two main ways of measuring distance in astronomy. The first can be used for nearby stars (that is stars which are not more than a few hundred light-years* away from the Sun). If you hold up a finger before your eyes and look at it first with your left eye and then with your right you will see that it seems to oscillate backwards and forwards against the background. The amount of the oscillation depends on the distance of the finger from the eye. This is called the *parallax* effect. As the Earth goes round the Sun each year, its positions in, say, mid-summer and mid-winter can be thought of as equivalent to the right and left eyes. A nearby star may then take the role of the finger held before the eyes. In just the same way, it will oscillate backwards and forwards against the background of more distant stars. In this case, too, the amount of oscillation will depend on the distance of the star. Hence the distance can be estimated by measuring the oscillation. A star 300 light-years away gives such a small oscillation that it is virtually indetectable. This therefore represents an upper limit beyond which measurement of distance by parallax methods is impracticable.

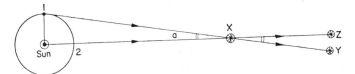

Fig. 1. Stellar parallax. As the Earth moves from point 1 to point 2 in its orbit round the Sun, star X appears to shift from being in front of star Y to being in front of star Z. The angle 'a' is the parallax of star X.

Once the distance to a star has been established, it is a relatively straightforward task to measure its apparent brightness. The actual, intrinsic brightness of a star depends in a very simple way on distance and apparent brightness, and so can be derived at once.

The other main approach to distance determination in astronomy is aimed at those stars (and they form an overwhelming

* The light-year — the distance light travels in a year — is a convenient unit in astronomy. It is equal to about 6,000,000,000,000 miles.

majority) that are more than 300 light-years from the Sun. In this, a specific type of star is chosen: preferably one that can be easily recognized. The Sun's immediate neighbourhood is then scoured for examples of this type of star and, when they are found their intrinsic brightnesses are determined — perhaps by the parallax method. It is then assumed that all stars of this type have the same brightness; or, maybe, their brightness depends in a simple way on their characteristics. Then, whenever a star of this type is met with at greater distances, a comparison of its apparent brightness with its assumed intrinsic brightness tells us at once how far away it is. By searching out these stars we can eventually build up a map of the star system around us. It is as if we were trying to draw a map of a town, being restricted to what we could see from the roof of one house. We find that we can plot out the shape of the streets but we cannot determine the scale. Then we discover a method of measuring the distances to some of the street lamps. When we plot the positions of these street lamps on our map, then we can find how far away are the houses and streets we can see.

We have been stressing the difference between the *apparent* brightness of a star — which is what we actually measure — and its intrinsic or *absolute* brightness. Even when differences in distance have been eliminated, so that we are dealing only with absolute brightness, we find that the stars still differ tremendously from one another in their brightness. The brightest star known is some hundred thousand million times more luminous than the faintest. This range is so great that astronomers find it easier to refer to stars in terms of their *magnitude* rather than their brightness. One star is a magnitude brighter than another if it is about two and a half times more luminous. Using this new term, we can say that the difference in brightness between the brightest and faintest stars observed is 28 magnitudes (which is obviously a much easier quantity to handle).

So far, we have been talking about the brightness of a star as it appears to our eyes. But stars may emit much of their energy in forms that we cannot see: as X-rays, ultra-violet light and so on.

If we want to talk about the total energy that a star emits (which is what astronomers mean by luminosity) then we must measure these invisible forms of energy too. Here, however, we run into trouble. The atmosphere of the Earth is only transparent to light and radio energy; all other forms are absorbed. They can only be detected by instruments which are lifted above the atmosphere: either in rockets or artificial satellites. Many measurements of this type have now been made, so that we currently have knowledge of stellar luminosities over much of the energy range of stars.

Surface Temperatures, Colours and Spectra

Suppose we do an experiment. We take a poker and heat the end of it in a furnace. The end, when we take it out, is white-hot. As it cools down the colour changes, becoming red-hot, till finally it radiates no light at all. If you now look at the stars through a telescope, or even a pair of binoculars, you will soon observe that they too differ among themselves in colour. Some are bluish-white, some red, and some are colours in between. These colour differences, as for the poker, reflect differences in temperature. We can therefore deduce immediately that blue or white stars have hotter surfaces than red stars. (Notice that this deduction applies only to the surface — the part we see. We cannot say immediately what the temperature below the surface is likely to be.) The colour of a star is usually obtained by inspecting it through two differently coloured filters — a blue and a red, for

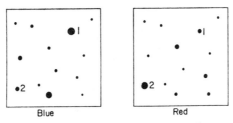

Blue   Red

Fɪɢ. 2. A star field photographed successively in blue light and in red light. Star 1 is a typical blue star; star 2 is a typical red star.

example. Then a red star will appear very bright through the red filter, but faint through the blue filter, whereas the opposite will be true for a blue star. The inspection nowadays is not made with the human eye, but on photographic plates or with photoelectric detectors.

There is another, rather more complicated way of obtaining stellar temperatures. This depends on the *spectrum* of the starlight. If you look at an electric light through a glass prism, you will be able to see the filament as a series of coloured images overlapping each other. That is, you have obtained a spectrum of the filament. If you put a prism at the end of a telescope and turn this combination until it points at a star, you will find, in the same way, that the image of the star is drawn out into a coloured line, which goes from red at one end, through yellow, green and blue to violet at the other end. This is a stellar spectrum. Normally, the light from a star is made to pass through a narrow slit before it strikes the prism. In this case, the starlight is drawn out into a coloured band rather than a line. If this band is examined in detail, it will be found to have a large number of black lines lying across it. White light is a mixture of light of all different wavelengths.* Spreading white light out into a spectrum is equivalent to decomposing it into its component wavelengths. Black lines in the spectrum therefore correspond to wavelengths where there is light missing.

The absorption (or emission) of light at specific wavelengths is a property peculiar to atoms and molecules. If we think of the simplest atom — hydrogen — we can picture it as a massive particle with a positive electrical charge (the *proton*) round which moves a light particle with a negative electrical charge (the *electron*). We might compare it with the Earth moving round the Sun. The main difference is that the Earth could circle the Sun

---

*Light is an oscillation, like waves on the surface of the sea. The colour of a given type of light depends on the distance between the crests of successive waves. This is the wavelength. The shortest waves to which our eyes are sensitive have wavelengths of about a hundred-thousandth of an inch, and appear violet to us. On the long wavelength side, the light appears to be red and has a wavelength twice as long.

at any distance so long as it had the right speed and direction, but an electron can only circle the central proton (or *nucleus* for a more complex atom than hydrogen) at certain specific distances. Whenever the electron jumps from one of its allowed orbits to another, it either absorbs or emits radiation (sometimes light, sometimes X-ray, sometimes radio: depending on the size of the jump). A jump outwards, away from the nucleus, corresponds to absorption of radiation; a jump inwards corresponds to emission. As the radiation from the interior of a star streams outward through the stellar surface, the atoms in the surface region absorb some of the flow. This produces black absorption lines in the spectrum.

The ability of an atom to absorb light depends on the temperature of its surroundings. For any particular type of atom there is a temperature at which it absorbs most efficiently. At higher temperatures it may begin to break up and lose electrons (become ionized); at lower temperatures it may not be sufficiently excited to jump at all. By studying the absorption lines in a stellar spectrum, we can therefore determine the surface temperature of a star. Fortunately, the temperatures obtained from the colours and the spectra of stars agree quite well with one another. The range of surface temperatures is relatively small; the hottest stars known have surface temperatures in the range 50,000–100,000°C, whilst the coolest may be at about 3000°C. (Recently, stars whose surface temperatures may be about 1000°C have been discovered.) Thus surface temperatures may vary by a factor of a hundred, which is much less than the variation in stellar brightnesses.

Since any given atom produces absorption lines most efficiently at a specific temperature, and since this temperature varies with the type of atom, it follows that the appearance of stellar spectra changes considerably as the temperature is altered. The temperature is indeed estimated from the overall change in appearance. It has been found convenient to group stars together on the basis of their spectra: each group therefore covering a certain range of temperatures. The groups are identified by letters of the alphabet (not arranged in any particular sequence). This method of identification is called the Harvard classification (for it was first proposed at that University). The table below relates this classification to the temperature ranges involved. It is noticeable that the hotter stars (they are often called *early-type* stars) cover a

Harvard Classification	O	B	A	F	G	K	M
Temperature Range (°C)	50,000–25,000	25,000–11,000	11,000–7,500	7,500–6,000	6,000–5,000	5,000–3,500	3,500–

much larger range of surface temperatures than the cooler (or *later-type*) stars. We find, in fact, that the Harvard sequence of spectral types (and the sequence of stellar colours too) does not depend directly on the temperature, but rather on the logarithm of the temperature.

FIG. 3. Examples of stellar spectra.
(a) O-type star   (b) A-type star
(c) G-type star   (d) M-type star

Sizes of Stars

Just as stars may differ considerably in brightness and surface temperature, so they may differ considerably in size. Data on sizes are, however, very difficult to come by. All stars appear as points of light, even in the largest telescopes: their diameters can only be measured under special circumstances.

One particularly useful method of studying sizes is by the examination of double stars. Double stars (often called *binaries* by astronomers) consist of two stars moving round each other: in much the same way that the Earth moves round the Sun. The two stars are called the components of the binary, and the path they follow relative to each other is called their orbit. Some double star orbits are orientated so that they are edgeways on to the Earth. As the two components go round, they pass alternately in front of each other. So, to an observer on Earth, they block out each others light in turn. A double star of this type is called an *eclipsing* binary. (An eclipse in astronomy occurs whenever one body blocks out the light from another. Thus solar eclipses occur when the Moon blocks out the light we receive from the Sun.) The length of time the eclipses last depends on the relative sizes of the two stars, and also on the relative speeds with which they move.

The components of eclipsing binaries are too close together in the sky for them to be seen separately through a telescope. The spectrum of such a double star therefore contains black absorption lines contributed by both members. (Although, if one star is much brighter than the other, its excess light will swamp out the lines due to the second star.) Now the exact wavelength of a stellar absorption line depends on the way in which the star is moving relative to the Earth. If the star is coming towards us, the lines are shifted slightly towards the violet end of the spectrum. If it is receding, they are shifted towards the red. If it has no motion towards us or away (that is, if it is either stationary or moving across our line-of-sight), then the lines are in their normal positions. This change in wavelength depending on the motion is called the *Doppler effect* (after its discoverer). The best-known example of the effect is actually in the field of sound rather than light. A train approaches from the distance blowing its whistle. At the start, the pitch sounds higher than usual. As the train comes closer and passes by the note drops until it reaches its normal pitch. Then, as the train recedes into the distance, the pitch drops further so that it is below the normal level. These variations are a result of the Doppler effect changing the apparent wavelength of the sound, and so altering the pitch.

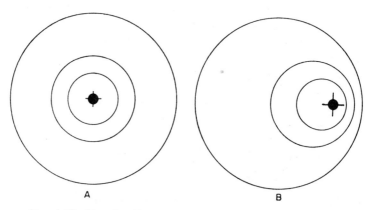

FIG. 4. The Doppler effect. In A the source of the waves is station-ary, so the wavelength is constant in all directions. In B the source is moving from left to right. If the source approaches an observer, he notes a decreased wavelength; if it recedes from him, he notes an increased wavelength.

Suppose we now apply this concept of the Doppler effect to an eclipsing binary star. As one of the components swings round its orbit, it first approaches the Earth, then it crosses our line-of-sight, then it recedes and, finally, it crosses our line-of-sight again but in the opposite direction. Correspondingly, the stellar absorp-tion lines first move towards the violet end of the spectrum, then they return to their normal position, then they move towards the red, then back again. This cycle of events is, of course, repeated over and over again. Meanwhile the other component of the binary is going through the same sequence of motions, but in the opposite order. When the first component is approaching the Earth, the second is receding, and so on. This means that the absorption lines of the two stars also always move in opposite directions. If we were to watch the spectrum, we would see the two sets of lines oscillating backwards and forwards across each other. The size of the oscillation depends on the relative speeds of the stars. (In our example of the railway train, it requires a high-speed express to produce a really good result.) The faster the speed of the stars, the greater the oscillation of their spectral lines. This means that, if we measure the amount of the oscillation, we can determine the speeds of the stars.

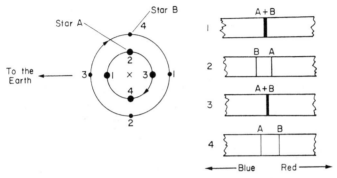

Fig. 5. The Doppler effect in an eclipsing binary.

After this long diversion, we can now go back to our initial problem about the sizes of the stars. We have said that the length of time an eclipse lasts depends on the sizes of the objects involved and the speeds with which they are moving. We can observe how long the light of each star in an eclipsing binary is blocked out by the other. We can also determine the speeds of the stars from looking at their spectra. Therefore we can calculate the sizes of the two stars.

There is another, more direct method of measuring stellar sizes. It is, however, difficult to apply, and is, moreover, restricted in its application to fairly large, nearby stars. It employs a stellar interferometer. Light, we have seen, is a wave-like motion. Waves on the surface of the sea often meet other waves going in other directions. When they do, *interference* occurs. If, for example, the trough of one wave coincides with the crest of another, the two cancel each other out (or interfere). If, on the other hand, two crests coincide, they add together to form a wave twice as big. The same sort of thing can happen with light. In particular, the light coming from the opposite edges of a star can interfere. The stellar interferometer is an instrument that detects this interference and so infers the size of the star. (Actually, to derive a size for the star by this method also requires a knowledge of its distance, which may not be easy to obtain.) So far as they go, these measurements confirm the sizes derived from studying eclipsing binaries.

The results show that some stars are several hundred times larger than the Sun, but others are less than a hundredth of its size. Astronomers have given names to the different groups of stars depending on their characteristics, and size is one of the criteria used in this naming. So, for example, stars that are ten, or a hundred times the size of the Sun are called *giants*. Still bigger stars are called *supergiants*. The size of a star, its temperature and its brightness are all related. The larger the star, the greater its surface area, and therefore the larger the region that can pour energy into space. Similarly, stars with high surface temperatures emit more energy than comparable stars with low surface temperatures. So a large, hot star will be extremely bright and a small, cool star extremely dim. On the other hand, a large, cool star and a small, hot star may both emit the same amount of energy.

Hertzsprung-Russell Diagram

Brightness and surface temperature are the two stellar characteristics that an astronomer finds it easiest to measure. They are therefore used extensively as a way of differentiating one star from another. The most informative way of showing such differences is by means of a diagram. We draw a graph showing values of the stellar magnitude vertically and values of, say, the colour horizontally. (It should be noted that the magnitude of a star depends on the logarithm of its brightness, and the colour — or spectral type — of a star depends on the logarithm of its surface temperature. So that equal distances across the graph represent equal *ratios* of the quantities concerned.) The surface temperature on the diagram, contrary to normal scientific practice, increases from right to left. Such a graph is called a Hertzsprung-Russell diagram (after its originators). This is often abbreviated to H-R diagram.

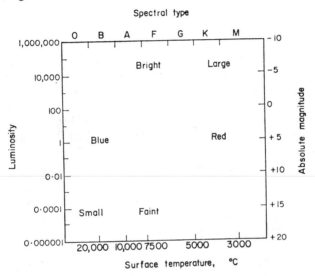

Fig. 6. The Hertzsprung-Russell diagram. (The luminosity is given as a fraction of the Sun's luminosity.)

We have seen that the brightness and surface temperature of a star are linked to its size. This means that the H-R diagram also gives some indication of size. The relationship is shown in Fig. 6. It can be seen that the size of a star increases with distance from the bottom left-hand corner of the diagram. It is often possible to divide stars into different types according to their position in the diagram. For example, stars towards the top right-hand corner of the diagram are obviously large in size. They have low surface temperatures and are therefore called *red giants*. On the other hand, a star down in the bottom left-hand would be very small; it would also be hot. Such a star is called a *white dwarf*. It is found that the vast majority of stars, when plotted in the H-R diagram, lie along a diagonal line running from the bottom right-hand corner to the top left-hand corner. This zone has therefore been called the *main sequence*. The Sun, itself, is a member of this sequence, lying on it at about the mid-way point.

Stellar Masses

Eclipsing binaries are one of our major sources of data on stellar sizes. They also represent our best method of weighing a star to find out how massive it is. We have talked so far of two stars circling round each other without explaining why they should do so. In fact, they are kept together by the same force that keeps the Earth circling round the Sun — their mutual gravitational attraction. Now this force depends only on the distance apart of the two stars and their masses. It is easy enough to find the distance between the two stars in an eclipsing binary. We can determine from their spectra how fast they are moving, and we can observe directly how long it takes them to complete a circuit. (This time — the *period* of the binary — can be found from the length of time between eclipses.) These two measurements together enable us to determine the size of the orbit. Knowing this, we can finally find the masses of the stars. The main drawback of this method is that the two components of an eclipsing double star are often only a short distance apart. Their gravitational interaction is then very strong: so strong, indeed, that they may pull material off each others surfaces. This may complicate the measurements so much as to make them valueless.

There is, however, another type of double star that can be used to determine mass. This is one whose components are so far apart that they can be distinguished as separate points of light through a telescope. A system of this type is called a *visual* binary. Here again, the masses of the stars can be found once the size of the orbit and the period are known. There is still another complication in this case. Since a visual binary can have its orbit orientated at any angle relative to an observer on the Earth, we do not know the true shape of its orbit immediately. We must first determine the inclination of the orbit before the masses can be found.

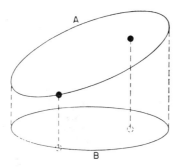

FIG. 7. A is the true orbit of the visual binary; B is the apparent orbit as projected onto the celestial sphere.

Components of visual binaries, being far apart, are never distorted by excessive gravitational interaction: unlike the eclipsing binaries. They nevertheless have disadvantages of their own. In particular, stars which are very far apart naturally take a very long time to complete one orbit. As a result, whereas the average eclipsing binary may have a period of a few days, the average period for a visual binary is a considerable number of years. There are, indeed, very many visual binaries with periods so long that not even one circuit of the stars has yet been accomplished since their discovery.

These various difficulties mean that really accurate stellar masses are very hard to obtain. There are not more than a few dozen altogether, and these are almost entirely for main-sequence stars. No supergiant masses are known at all. If the available masses of main-sequence stars are plotted in a diagram against their magnitudes it is found that they lie more or less along a line. This line represents a *mass–luminosity* relationship. Since the main-sequence band in the H-R diagram represents a surface temperature – luminosity relationship, we can see that the mass, luminosity and surface temperature of main-sequence stars are all inter-related.

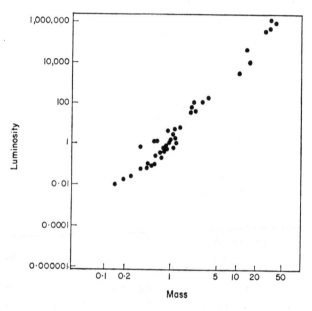

FIG. 8. The Mass–Luminosity relation. Both mass and luminosity have been expressed as ratios relative to the Sun. Note that both scales are logarithmic.

Chemical Composition

There is one major stellar characteristic still left for us to discuss. This is the chemical composition of the stars. The questions we must ask are: What chemical elements do stars contain? In what relative proportions? Do these proportions vary from star to star? It is only in the last twenty years that answers to these questions have begun to appear. To determine how much of a given element there is in a star, we must rely on measuring the amount of light that that element absorbs on the stellar surface. But this depends on many other factors besides the relative abundance of the elements present. As we have seen, it will depend initially on the conditions within the stellar atmosphere — the temperature and pressure. Therefore, before we can discuss the chemical composition of a star, we must first eliminate a large number of other effects. Even if this is done with very great care, the possible error in the results can be large. The best measurements may differ by as much as a factor three amongst themselves. Fortunately, differences in the abundances of elements may vary by much more than this from star to star, so that, despite the inaccuracy of the measurements, it is still possible to obtain positive results. Indeed, variations in the amount of the elements present can be so large as to produce differences in the apparent colour of otherwise similar stars. This arises because the absorption lines remove more light in the violet region of the spectrum than in the red.

The most obvious feature of the chemical abundances in different stars is the surprisingly high degree of uniformity. In order to find really striking differences of composition it is necessary to examine quite a large number of stars. Another major feature is the great preponderance of hydrogen and helium. A typical star like the Sun might have nearly three-quarters of its mass in the form of hydrogen, and most of the remainder as helium: all the other elements together only contribute some 2 per cent of the total mass. If we remember that hydrogen and helium are the lightest of the elements, the abundances expressed in terms of the number of atoms present are even more impressive. A sample of the Sun's atmosphere which contains a thousand hydrogen atoms also contains eighty helium atoms and not more than two other atoms. The Sun appears to be quite typical in these proportions.

Helium actually presents one of the major difficulties in the way of abundance determinations. It is most important to know whether the proportion of helium to hydrogen varies from star to star. Unfortunately, helium is a very poor producer of absorption lines. The result is that figures for helium abundances are much more inaccurate than for any of the other major elements.

Abundance measurements share with temperature measurements the disadvantage of referring only to the stellar surface. There is no direct way of determining whether or not the composition within a star differs from that at its surface. As we shall see, there are good theoretical reasons for supposing that it does. Indeed, theory suggests that stellar evolution is bound up with gradual changes in the chemical composition of stars. These take place predominantly in the stellar interior, so that we only rarely see the results. Sometimes, however, evidence of these changes does appear and can be discovered by making abundance comparisons. The most obvious example of such evidence is the element technetium. This is extremely rare on the Earth; it was, in fact, first produced artificially in 1937. It is scarce because it is radioactive, and so can decay away fairly rapidly on an astronomical time-scale leaving only negligible amounts. Yet this same element has been seen to absorb light quite strongly in the atmospheres of certain giant stars (they belong to a special grouping, designated by the letter S, in the Harvard spectral classification). How has technetium managed to survive in these stars but not on the Earth? We must suppose that the element is actually being created in the stars — as it has been in terrestrial laboratories — so that we are really seeing here a star which is changing its chemical composition. Since one element can only be produced by transmuting another, we must suppose that the increase in the amount of technetium present is balanced by a decrease in some other element.

Minor Characteristics

We have now described the main stellar characteristics which it is possible to measure directly. There are a few other characteristics which it would be useful to know, but which either cannot be measured accurately or which can only be measured for a very few stars. For example, it would be helpful to know just how the material within a star is distributed. Are stars very dense near their centres and very rarified near their surfaces, or does the density change in a regular, gradual manner throughout the star? This question can be answered, to some extent, from a study of very close eclipsing binaries. We have already noted that such stars exert an excessive gravitational pull on each other so that they become considerably distorted. The exact nature of the distortions depends on the way in which the material is distributed within the stars. On the other hand, the distortions of the two stars gradually produce a change in their orbits. So, if the orbits are studied in sufficient detail, we can hope to find out something worthwhile concerning the density distributions in the stars. The method is only of restricted use, however, because there are very few close binaries which are suitable for measurement.

Ages of Stars

There is one piece of information about stars that astronomers would give a good deal to know, but are completely unable to measure directly. This is their age. It is certain that, however stars evolve, the length of time since their birth must be a basic factor: just as it is in the growth of a plant or animal. But, whereas you can follow the development of a plant or animal from its birth to its death, it is impossible to follow similarly the development of a star. In terms of astronomical change, a human lifetime is nothing. The whole of recorded history is too small a unit. We must think in terms of millions of years, even of thousands of millions of years. Sir William Herschel, a hundred and fifty years ago, could already point out this prime difficulty in the study of the stars. He compared the astronomer with a man who, never having seen a tree in his life, is allowed to walk for an hour through a forest. During that time he would not see a single change take place in any given tree. But he would see seedlings, young trees and mature trees, and he would see decaying and lifeless tree-trunks. If he were acute enough to piece the evidence together, he would be able from his brief excursion to work out the life-history of a tree. This sums up accurately the relationship between our observations of the stars and our theories of how they evolve. The former apprise us of the current situation; the latter try to explain it.

To some extent this description of our knowledge of stellar ages is too gloomy. As we shall see, it is possible to make one or two indirect checks on age. These form very useful occasional tests of our theories of stellar evolution.

The Sun

So far we have completely ignored the fact that we live very close to one particular star — our Sun. Much of the information which is difficult to obtain for other stars is easily available for the Sun. We know the distance to the Sun very accurately, and therefore we can determine its size and the amount of energy it emits with considerable precision. The Sun is the only star whose surface we can study in detail. This means that we can derive much more than just an average surface temperature: we can also decide whether the temperature varies across the face of the Sun, or whether it changes with height above the solar surface. We therefore know much more about the atmosphere of the Sun than about any other stellar atmosphere. This, in turn, makes it possible to determine the chemical composition of the Sun to a higher degree of accuracy than for other stars. The mass of the Sun, too, is known very precisely from the gravitational pull of the Sun on the Earth.

It is even possible to find an age for the Sun. If we suppose that the Sun and the Earth were born together, then any age we can determine for the latter will also necessarily apply to the former. The Earth's crust contains certain radioactive elements which are gradually breaking down into others of less weight. If the rate of decay is known, and also how much has already broken down, then an age can be found for the Earth's crust. This, presumably, represents a lower limit to the age of the Earth as a whole. The radioactive element which has been most used for this purpose is uranium. It breaks down into lead and helium, both of which can be measured. Of recent years, other radioactive elements — potassium and rubidium, for example — have been used for age determinations. All these methods now give results in fairly good agreement. They indicate an age for the Earth's crust which is somewhat less than four thousand million years. It is usually assumed that the Earth as a whole (and therefore the Sun, too) is rather older, having an age of about four and a half thousand million years.

Because we know so much more about the Sun than about any other star, it is very convenient to use it as a touchstone for our ideas on evolution. If our theories can explain accurately the present characteristics of the Sun, we may apply them more confidently to other stars.

Summary

We have discussed in this chapter six stellar characteristics which we might reasonably expect to be important: luminosity, surface temperature, size, mass, chemical composition and age. Of these, luminosity and surface temperature are relatively easy to estimate. Size is rather more difficult but, since it is related to the first two characteristics, it can be obtained indirectly. Chemical composition is hard to determine accurately. Mass is impossible, except in special circumstances, and this is also true of age.

Which of these characteristics are basic in influencing the evolution of a star? As will become apparent in the course of this book, the stellar characteristics which determine evolution are also those which it is most difficult to measure — mass, chemical composition and age. Faced with this dilemma, the astronomer must tackle the problem of stellar evolution in a roundabout way. He must start by *assuming* a certain mass, chemical composition and age for a star. He then calculates what the corresponding luminosity, surface temperature and size should be theoretically. Finally, he compares his answers with the values of these quantities observed for any given star. If he is fortunate the two agree, and he then claims that he has calculated a *model* of the star. If, as is usually the case, he is less fortunate, he revises one, or all, of his original assumed values and tries again until he achieves a fit. It is because this whole business of repeated calculation is so tedious and time-consuming that great progress in the theory of stellar evolution has only been achieved in the last few decades. It awaited mainly on the development of high-speed electronic computers which could handle the computations with ease.

One query that might strike the cautious reader is this — How do we know that the models we have constructed are unique? Are astronomers so delighted when they manage to compute a stellar model fitting the requirements of mass, chemical composition, etc., that they never stop to consider whether another model, with a different internal structure, might not also fit these requirements? The answer is that astronomers are normally justified in their optimism. It can be shown — by an argument known as the Russell-Vogt theorem — that under almost all circumstances, there is only one stellar model that will fit all the observational requirements. Astronomers can therefore heave a sigh of relief as soon as they compute any self-consistent stellar model.

CHAPTER 2

Stellar Families

Initial Classification

We have considered the main observable properties of stars. Now we must consider how these properties can be used to separate the stars into types. This will form the second stage of our study of stellar evolution. Subsequent chapters will then discuss the inter-relationships which theory suggests exist between the various types.

It is usually easiest to differentiate between different sorts of stars by plotting their position in the Hertzsprung-Russell diagram. As we saw in Chapter 1, this separates stars out according to their brightness, surface temperature and size. The important feature of this diagram is that the stars are not distributed over it at random, but group together in certain regions. Figure 9 represents an H-R diagram with the most important of these regions shown. The different groupings in the H-R diagram are by no means equally populated with stars. The greatest number by far lie along the main sequence. The red giants form another fairly prominent group. There are a very few bright stars along the top of the diagram in the supergiant region, and a group of very faint white dwarfs at the bottom. Apart from these main groupings, there are a certain number of stars scattered about in other parts of the diagram. These are usually distinguished by their size. Stars on the main sequence are called dwarfs (although those at the upper end of the main sequence are distinctly larger than those at the lower end). Anything to the right and above the main sequence comes into the general category of giant stars. A star a hundred times bigger than the Sun is just called a giant. If it is

23

a thousand times bigger it is called a supergiant; if it is only ten times bigger it is a subgiant. In a similar way, a star which lies below the main sequence, but well above the white dwarfs, is called a subdwarf.

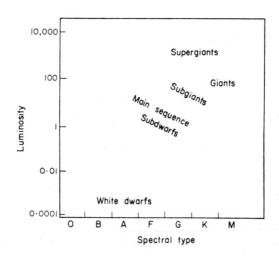

FIG. 9. Hertzsprung-Russell diagram showing main subdivisions of stellar types.

It is particularly important to remember that the number of stars plotted on an H-R diagram is often not representative of the actual numbers of the different types of stars in space. For example, supergiants, because they are so bright, can be detected very easily — they have been observed over a very large volume of space round the Earth. White dwarfs, on the other hand, being very faint, are extremely liable to be overlooked. They have only been detected in the region of space relatively near the Earth. The difference can be seen quite plainly if we compare the H-R diagram for the stars which appear brightest to us on Earth with the H-R diagram for the nearest stars (Figs. 10 and 11). The first contains several giant stars, but no white dwarfs. The second

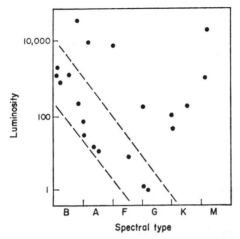

Fig. 10. The Hertzsprung-Russell diagram for the stars of greatest apparent brightness. The stars within the dotted lines form the upper end of the main sequence. The remaining stars are giants.

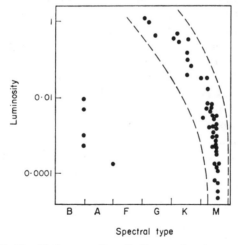

Fig. 11. The Hertzsprung-Russell diagram for the nearest stars (within 15 light-years of the Sun). The stars within the dotted lines form the lower end of the main sequence. The remaining stars are white dwarfs.

contains several white dwarfs but no giants. It is obvious that the latter diagram gives a truer picture of what the space around us is like. Very bright stars are also very rare; very faint stars are very common.

As we have seen, stars can also be plotted on a mass–luminosity diagram. Unfortunately, the data are too few for stars other than main–sequence stars, so we cannot determine what kind of groups they form in this new diagram. However, the two diagrams together show definitely that only certain values of brightness, size, surface temperature and mass can be found in a star: these four different characteristics cannot be chosen at random — they are interdependent.

Physical Groupings of Stars

Ideally, in our attempt to group stars together into families we would like to examine all the stars in the Universe. In practice, of course, this is impossible: apart from anything else, the number of stars involved is far too large. What we must do, if possible, is to look for actual physical groupings of stars in the sky. We can then isolate one of these smaller regions of the Universe and try to examine a good-sized sample of its stellar contents. (This is equivalent to the practice in botany of taking a sample plot of land, counting the different sorts of plants present, and assuming that the result represents the average for all areas of the same type.) Fortunately, the stars do come together in large natural units which may reasonably be studied as separate entities. These units are the galaxies; a large galaxy may contain some hundred thousand million stars. Our own Sun lies towards the edge of one such large galaxy. (We call it, rather grandiosely, the Galaxy.) From our vantage point on the outskirts, we must try to examine the stars within our Galaxy, and to separate out the different types and families. Then, if our telescopes will permit it, we must examine other galaxies and see whether they contain similar, or different stellar families.

The Galaxy

Our own Galaxy is a spiral. That is to say it consists of a large number of stars forming a central nucleus out of which spring several spiral arms. These arms contain stars, as the central nucleus does, but they also contain clouds of gas and dust which are lacking in the nucleus. There is another immediately obvious difference: the brightest stars in the nucleus are red, which gives the nucleus a decidedly reddish tinge, but the brightest stars in the spiral arms are blue, so the arms have a generally bluish appearance. Moreover, the whole assemblage of nucleus and arms is considerably flattened: a cross-section through the Galaxy bears some resemblance to the cross-section through a discus. As a result, when we look out at night, we see the cross-section of our Galaxy as a relatively narrow band circling the heavens. This is the Milky Way.

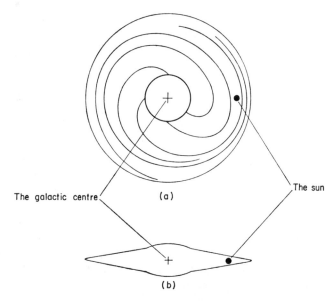

FIG. 12. (a) Plan-view and (b) cross-section of our Galaxy.

The clouds of gas and dust (the *interstellar* material) are of considerable interest for stellar evolution, but they are a great nuisance in practice. The trouble is that the dust hinders our observations of the stars. The astronomer is in rather the same position as a man in a foggy street: he can see the street lamps that are near at hand, but the ones that are further away are completely blotted out by the fog. This means that we cannot study a random sample of the stars in our Galaxy as we would like. Instead, we must study those that are visible and then try to deduce from this whether we have missed anything.

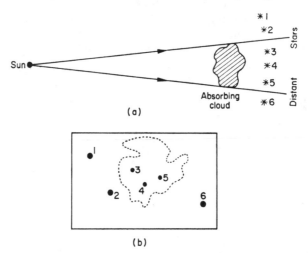

FIG. 13. The effect of dust on the observation of stars: (a) shows six stars of equal brightness, three of which are situated behind a dust cloud; (b) shows the resulting appearance on a photographic plate.

Star Clusters

When we examine the stars in our Galaxy, we find that many of them are not wandering about individually like the Sun, but are grouped together into clusters. Such clusters are particularly important in the study of stellar evolution. We have seen that one of the important stellar characteristics is age, but that age cannot be measured directly. It seems certain that stars which now appear together as a group have always been together since their birth. We may also suppose, as a first approximation, that they were born simultaneously, so that they all have the same age. Thus, though we may not be able to measure just how old a cluster of stars is, we can at least say that the differences between stars in a cluster are not due to differences in age. In other words, by studying stars in a cluster we can eliminate one of the unknown factors from our work.

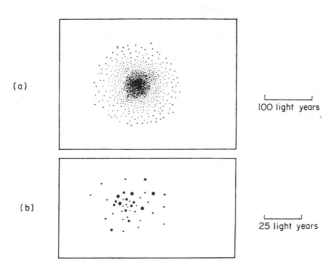

(a)

100 light years

(b)

25 light years

FIG. 14. (a) A globular cluster. (b) An open cluster.

Star clusters can be divided into two types: globular clusters and galactic, or open, clusters. The first type are large — a big one may contain several hundred thousand stars — whereas the latter do not usually consist of more than a few hundred stars. Moreover, the stars in a globular cluster are somewhat closer together than those in an open cluster. It is these differences which give rise to the characteristic forms of the two types of cluster: the globulars have so many stars at their centres that they look like luminous balls, whilst the open clusters have each of their stars individually distinguishable. There are other differences of equal importance, although sometimes less obvious to inspection. The galactic clusters often contain a good deal of gas and dust lying about between the stars; the globular clusters do not. The brightest stars in globular clusters are red in colour (large and cool); the brightest stars in many galactic clusters are blue (smaller but hotter). When the whole set of clusters visible from the Earth is examined, it is found that the two types of cluster also differ in their location in the Galaxy. The open clusters — several thousand of them — are situated in the spiral arms, with which they obviously have much in common. The globular clusters — only about a hundred are known — congregate round the nucleus of our Galaxy to which they obviously bear a resemblance. (Since the Sun lies towards one edge of the Galaxy, most of the globular clusters seem to us to be packed together in one part of the sky.)

The difference between the globular clusters and the nucleus of our Galaxy on the one hand, and the open clusters and the spiral arms on the other, suggests that the stars they contain may fall into different families. We can examine this possibility first of all by constructing H-R diagrams for the two groups and looking for differences. We start by examining the H-R diagrams for the clusters. This is simpler than looking at the stars which are moving independently in the arms or the nucleus (they are usually called *field stars*), for we know that every cluster has a definite age, whereas field stars may have vastly differing ages. It is, in any case, very difficult to observe the nucleus, for the large quantities of dust in the Milky Way almost blot it out. This is one of the penalties we must pay for living towards the edge of the Galaxy. Once we have constructed H-R diagrams for a number of clusters, we can compare them with diagrams for various samples of field stars and see whether there are any differences.

If we compare the H-R diagram of a typical open cluster with the diagram for field stars near the Sun, we find that there is an overall similarity. The cluster diagram is, however, more clearly defined. The main sequence is narrower, and there is an evident gap between the top of the main sequence (the bright, blue stars) and the red giants present. (This is called the Hertzsprung gap.) In some open clusters, the main sequence at its lower end lies above the main sequence for nearby field stars. The H-R diagram

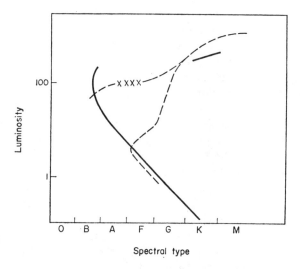

FIG. 15. (a) ————— Hertzsprung-Russell diagram for a typical open cluster. (b) - - - - - - - Hertzsprung-Russell diagram for a typical globular cluster.

for a typical globular cluster, on the other hand, is quite different. The main sequence is very hard to discover; it consists of a short band at the bottom of the diagram. At its upper end this band deviates to the right and curves upward into the red-giant zone. (There is no Hertzsprung gap.) From the red-giant region another band crosses the diagram horizontally. This *horizontal branch* therefore consists of fairly bright stars, all of the same brightness but covering a range of colours. In the middle of this branch (marked by crosses in Fig. 15) occurs a group of stars which vary appreciably in brightness and colour over a period of less than a day. They are known as RR Lyrae stars.* So far as can be judged from the rather meagre data on nucleus stars available, the H-R diagram for the nucleus of our Galaxy resembles a typical globular cluster diagram.

Stellar Populations

Our examination of H-R diagrams indicates that there are two groups of stars which differ both in their positions in the Galaxy and in their intrinsic properties. These groups have been given distinguishing names. Stars in open clusters and related field stars are said to belong to Population I; stars in globular clusters or of a similar type belong to Population II.

In fact, the differentiation is not as clear-cut as this division would suggest. Clusters have been found which are, to some extent, intermediate in their properties. For example, there is a well-known open cluster designated as M67. (This means that it is the sixty-seventh object in a catalogue published by the French astronomer, Charles Messier, in 1781.) At first sight, its only peculiarity is that it has a considerably greater number of stars than the average open cluster. Its H-R diagram, however, reveals a different story. As Fig. 16 shows, the stars in M67 form a continuous band up to the right from the main sequence — just like a globular cluster. On the other hand, there is no sign of the typical globular cluster horizontal branch.

If we can make an initial separation of clusters into two stellar populations, can we do the same for field stars? The H-R diagram for field stars in our local region of the Galaxy leads us to suspect right away that they do not form a homogeneous group. For one thing, some stars occupy unexpected positions in the diagram; for another, the main sequence seems wider and less well-defined

---

*Stars which vary in brightness are called either after the brightest star of that type known, or after the first discovered. The combination of letters of the alphabet with the constellation name is a typical way of denoting variable stars.

than for clusters. This does not necessarily mean, of course, that **the** space around us is occupied by both stellar populations — there may well be other explanations. It does mean that we

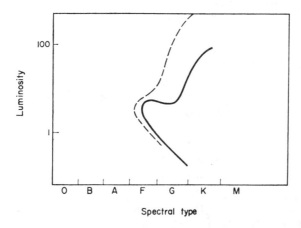

FIG. 16. The Hertzsprung-Russell diagram for M67 (the corresponding part of the diagram for a typical globular cluster is shown for comparison.)

cannot say confidently that the spiral arms of our Galaxy are similar to open clusters and therefore contain only Population I stars. It may be that there are Population II stars present as well. We must investigate further.

Chemical Composition and Stellar Populations

We have set up two stellar populations on the basis of the H-R diagram: that is in terms of the brightness, surface temperature and size of a star. Now we will consider this division in terms of another fundamental characteristic — the chemical composition. If the spectra of stars in open clusters (Population I) are compared with those of stars in globular clusters (Population II) an immediate difference appears. The lines of all the elements **except** hydrogen and helium are appreciably stronger in **the** former than in the latter. An estimate of this difference shows that a typical Population I star has about ten times more of the heavier elements than a Population II star. ('Heavier' here means 'elements heavier than hydrogen and helium'.) Despite this considerable difference, we must remember that typical stars of both populations are predominantly — over 95 per cent — made up of hydrogen and helium. If we extend our examination over several clusters, we find that the exact proportion of the heavier elements present can vary. For the chemical composition, as for the H-R diagrams, there is not just a single division into two completely separate groups.

Stellar Motions and Populations

If we extend our examination of the chemical composition to field stars in our local region, we find the same range of values as for all the clusters — from open to globular. This is significant: it indicates that there are field stars in the spiral arms with properties similar to Population II stars. But a much more important fact than this also becomes evident — differences in the chemical composition of stars are related to differences in the way they move.

The Galaxy as a whole is rotating. In particular, the stars in the spiral arms are spinning round the central nucleus. Our Sun moves round the nucleus in much the same way that the Earth moves round the Sun. The Earth actually follows a fairly circular path round the Sun, but some objects in the solar system (comets, for example) prefer very elongated paths: dipping in close to the Sun at one end of their orbit and retreating far into the depths of space at the other. If we examine the motions of the stars in the neighbourhood of the Sun, we find that some are following fairly circular paths round the galactic nucleus whilst others are following highly elongated orbits which dip well in towards the centre of the Galaxy (Fig. 17). It is found that these different possibilities for the orbit actually correspond to our previous distinction in terms of chemical composition. A star whose chemical composition is typical of a Population I star also follows a fairly circular orbit round the nucleus; one whose chemical composition corresponds to Population II follows an elongated orbit which approaches the nucleus at its inner end. We are, in fact, being led back to our first distinction between population types on the basis of their position in the Galaxy. The observations suggest that the spiral arms contain both Population I and Population II stars (though the former more than the latter), but the Population II stars are still linked with the central regions

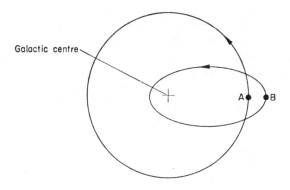

FIG. 17. Orbits of typical Population I (A) and Population II (B) stars round the centre of the Galaxy.

of our Galaxy. Again there are many intermediate stars, whose orbits are neither circular nor highly elongated. These are also the stars which have intermediate chemical compositions. Reverting momentarily to the clusters, we find that a study of their motions leads to the same results. The open clusters move round the nucleus in fairly circular orbits; the globular clusters move in elongated orbits.

In the solar system, the planets go round the Sun not only in fairly circular orbits, but also all more or less in the same plane. Comets, on the other hand, besides having elongated orbits also approach the Sun from all directions. The motions of Population I and Population II stars are directly analogous (Fig. 17). The former move round the galactic nucleus more or less in the plane of the Milky Way; the latter can rise to considerable heights above or below this plane. As a result, the proportion of Population II stars present per unit volume increases with distance above or below the Milky Way. Although near the Sun a majority of stars are Population I, if we move up or down from the Sun, vertically out of the plane of the Milky Way, within a few thousand light-years the Population II stars become the dominant component.

It was noted in Chapter 1 that variations in the amount of the heavier elements present could lead to apparent differences in the colour, or spectral type, of stars, even though they might have the same surface temperature. This means that H-R diagrams which are based on measurements of colour or surface temperature will, to some extent, separate out the Population I and the Population II stars. It is found, in fact, that Population II stars lie below the normal main sequence for nearby stars — they appear in the diagram as subdwarfs.

The Sun takes something like two hundred million years to go once round the Galaxy. This is a long time, but compared with the age of the Sun (nearly five thousand million years) it is relatively short. Many of the stars in the solar neighbourhood follow slightly different paths from the Sun. We can therefore deduce that they are only temporary neighbours of the Sun. During the course of one rotation round the galactic nucleus, they appear and disappear from our region of the sky. It is therefore reasonable to assume that the results we have obtained from examining our own local region will, in fact, apply throughout most of the spiral arms of our Galaxy.

Associations

Observations of the spiral arms have shown that they are not simply random collections of gas, dust and stars. There is a certain amount of organization present. The star clusters are one example of this. Another is that some field stars have been found to have related motions: they follow similar paths round the galactic nucleus. These particular streams of stars must therefore stay together over quite long periods of time. There are also localized regions (from ten to several hundred light-years in diameter) where the stars are evidently inter-related. These regions contain many bright, blue supergiant stars, together with an admixture of fainter stars and a great quantity of gas and dust. Among the fainter stars present is a particular type called the T Tauri (this designation indicating a variable brightness), which is especially notable since its position in the H-R diagram is slightly above the main sequence at the fainter, red end. (T Tauri stars are also observed in some galactic clusters.) It has been found, in fact, that two types of region can be distinguished: one where the supergiants are the most obvious feature and the other where the T Tauri stars are most evident. These are called respectively O-associations (because the brightest supergiants have a spectral class O) and T-associations (after T Tauri). These associations are distinguished from clusters by the greater volume of space they encompass and by the lower density of stars per unit volume.

It is possible that associations can provide one of the few ways of estimating stellar ages. Measurements of some O-associations

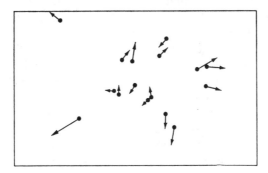

FIG. 18. The motions of stars in an O-association (the length of the arrows is proportional to the rate of motion of the stars).

seem to show that they are unstable: the stars are moving away from each other. We see them now some distance apart; formerly they must have been much closer together. If we trace this process of disintegration back in time, we reach a point when all the stars must have been clustered very closely together. Further than this it is impossible to go, so this point is taken to represent the time of their creation. Hence, from a knowledge of their present motions and distances apart, we can deduce an approximate value for their age. The only drawback is that the measurements of motion are difficult, and some observers deny that the results carry any conviction. However, it must be added that ages derived from the expansion of associations agree roughly with the ages of the same stars estimated by other methods.

The Spiral Arms

We have talked of star clusters, of associations and of streams of stars. All these are parts of the structure of a spiral arm. But we have not yet discussed what defines the spiral arm as a whole. How do the spiral arms differ from the spaces between the arms? They undoubtedly differ in the number of stars present: the spiral arms have appreciably more stars than the spaces between the arms. But the most obvious difference is not in their stellar content, but in the amount of gas and dust. It is concentrated almost entirely into the spiral arms. We may reasonably assume therefore that it is the presence of gas and dust which distinguishes the arms from the spaces in between.

Although stars may be detected in or out of spiral arms, it is found by observation that some types of stars — such as the bright, blue stars in O-associations — occur only in the arms, and they are usually very closely associated with clouds of gas and dust. Quite often one of these bright stars is close enough to the interstellar material to heat it very strongly. This causes the gas to emit spectral lines — like the absorption lines in the Sun's spectrum, but bright instead of dark. The appearance of these lines can be used to estimate the chemical composition of the interstellar gas. It is found in this way that it has much the same sort of chemical composition as the stars — predominantly hydrogen. (The dust does not emit spectral lines; but there is some evidence for its chemical composition from the infrared — 'heat' — radiation it emits.) There is, of course, no obvious reason why the gas and dust should always appear together. One could imagine having clouds made up of dust only in one part of a spiral arm, and clouds of gas only in another part. It seems much more probable from the observations, however, that the gas and dust do always occur together, although the relative proportion of the two may vary from cloud to cloud.

It is natural to want to trace the positions of the spiral arms in our Galaxy; but we come up against the old problem that the dust they contain blots out any distant view of the Galaxy. We can only plot directly those portions of the spiral arms which are fairly close to us. It is actually easier to find the positions of the spiral arms in other galaxies than in our own. However, a major development in radio astronomy at the beginning of the fifties has enabled us to overcome this difficulty.

It was predicted theoretically during the 1940's that hydrogen gas should emit a spectral line in the radio region at a wavelength of 21 centimetres (about $8\frac{1}{2}$ inches). This line was first detected by radio telescopes in the early 1950's. We know from our experience on Earth that radio waves can quite easily penetrate a fog which stops completely all the light. The same is true of the dust clouds in space. Whereas light is scattered by the interstellar material, the radio waves come through unchanged. This means that radio telescopes can detect 21-centimetre emission from hydrogen throughout the whole of our Galaxy. But we have seen that it is the gas clouds (predominantly made up of hydrogen) which define the positions of the spiral arms. Radio astronomy therefore allows us to map out the arms in our Galaxy even though we can never hope to see them all visually (Fig. 19).

If we compare the picture of our Galaxy derived from 21-centimetre observations with the visual observations of other galaxies, we find that our Galaxy seems to be a typical spiral galaxy. The significance of this for stellar evolution is that the results we derive from our own Galaxy are likely to have a much more universal application. We will take up this point again in the final chapter.

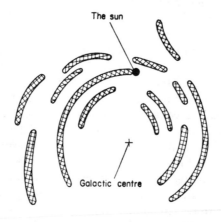

Fig. 19. The spiral arms of our Galaxy from 21-centimetre observations.

Variable Stars

In biology, it is often the freak specimens that cast light on the how and why of normal development. Similarly, in astronomy, the peculiar stars can often help us in our study of stellar evolution. 'Peculiar', in this case, means stars which differ noticeably in their characteristics from the main groups of stars (such as the main-sequence stars, or the red giants). One very common and obvious difference is that the brightness and surface temperature of a star may change. Stars which are peculiar in this way are called *variable* stars. Eclipsing binaries are usually included among the variable stars, since their light output appears to vary. We have seen, however, that the variations in this case are incidental: due to the tilt of the double star orbit. The sort of light variation we are now discussing is intrinsic in the star, and not due to some external factor.

The simplest way of classifying variable stars is by their light variation; whether it is periodic (repeating itself at regular intervals), irregular or even, as sometimes occurs, semi-regular (partly periodic and partly not). This classification is, of course, adopted purely for convenience: it does not indicate that all the members in each group are closely related.

Periodic variables have received the most attention; mainly because they have proved very useful in determining distances. The two most important groups are the *Cepheids* (named after the prototype star δ Cephei) and the *RR Lyrae* stars. Both of these are pulsating stars. The variation in their brightness is due to a corresponding variation in size — rather like a balloon that is periodically blown up and then deflated again. The Cepheids have periods ranging from one to fifty days. The RR Lyrae pulsate more rapidly, with periods of a few hours. The two types of star differ in their positions in the H-R diagram. The Cepheids are yellow supergiants; the RR Lyrae stars are blue giants.

Their use in measuring distances is also dissimilar. It has been found that the period of a Cepheid variable depends on its average intrinsic brightness. Now the period is a relatively easy quantity to measure, so we may readily determine the intrinsic brightness. A comparison with the average apparent brightness then gives the distance. Distance measurement with RR Lyrae stars is even easier. All RR Lyrae stars have a rather similar intrinsic brightness. This is apparent from the H-R diagram of a globular cluster (Fig. 15b). Once we know what this intrinsic brightness is, a comparison with the apparent brightness of any RR Lyrae star will immediately give its distance. The importance of knowing the distances either of Cepheids or of RR Lyrae stars is that they both occur in clusters. Once the distance to one of the stars in a cluster is known, the distance of the cluster as a whole is obviously known. Hence the luminosities and sizes of all the cluster stars can be determined.

Cepheids and RR Lyrae stars are both found as members of globular clusters. This suggests that they belong to Population II. It has been discovered, however, that Cepheids sometimes appear in open clusters as well. This apparent contradiction has been resolved by the realization that there are actually two types of Cepheid. One group (whose members are called *Type II Cepheids* or, sometimes, *W Virginis stars*) belongs to Population II and

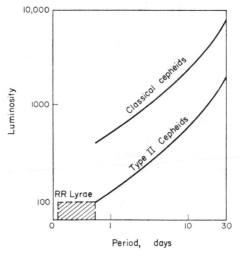

FIG. 20. Period-luminosity relationships for Cepheids and RR Lyrae stars.

therefore appears in globular clusters. The other contains the *classical Cepheids* which are Population I stars and may appear in open clusters. The most important difference between these two Cepheid groups is that they do not follow the same period — luminosity relationship. This means that the distance we measure depends on the type of Cepheid observed.

Besides the Cepheids and the RR Lyrae stars, many red giants and supergiants vary appreciably in brightness. All kinds of variability are observed in this group: regular, semi-regular or irregular. The regular variables are usually called long-period variables, since one cycle for these stars often takes months, or even years, to complete. The range in brightness can be very large: a star of this type may be ten thousand times brighter visually at maximum light than at minimum. The semi-regular and irregular stars generally change in brightness by a good deal less. The light variations in these giants and supergiants are undoubtedly sometimes due to pulsation. For example, the diameter of the red supergiant star Betelgeuse in Orion has been measured with an interferometer. Its light increases by about a factor of three between minimum and maximum. Correspondingly, the diameter changes from about $300R_\odot$ to $400R_\odot$.* In some of these stars, however, there is probably another factor at work. It seems likely that their atmospheres can become so cool at minimum light that some of the gas present condenses into solid particles. These form a shell effectively blocking much of the light from below. Then, as the temperature increases once more, the solid particles evaporate and the light again flows freely.

We have now covered the main groups of stars whose light variations can be linked in some way with pulsation. There is, however, another group of stars whose light variation derives from a different source. These are the *exploding* variables.

Exploding Stars

From time to time throughout history bright 'new' stars have flared up in the sky, only to disappear again after a few weeks or months. Two main groups of this type of variable have been distinguished — the *novae* and the *supernovae*. (As the names imply, the latter become much brighter at maximum than the former.) A nova, so far as we can tell, starts off as an intrinsically fairly faint star. It flares up very rapidly, increasing its brightness by a factor of sixty thousand or more. Then it declines again,

* The symbol $R_\odot$ denotes the radius of the Sun. Similarly, $M_\odot$ denotes the mass of the Sun, and so on.

more gradually, until, after a period of two years or so, it has returned to its original state.

Spectroscopic studies of novae have shown that the sudden increase in brightness corresponds to a stellar explosion which ejects the surface of the star into space. Sometimes, after the lapse of a few years, the ejected material can be seen directly through a telescope as an expanding cloud of gas round the star. The amount of the stellar surface thrown off depends on the size of the explosion, but an average value might be 1/10,000 of the total mass of the star. Some novae have been observed to flare up again after a few decades. Indeed, some nova-like stars are known which flare up perhaps once a month. It seems likely that all novae are potentially recurrent; the length of time between explosions depends on the size of the explosion — small explosions can occur frequently, large explosions only at long intervals.

If the brightness and surface temperature of a nova are determined when it is in a quiescent stage (either before or after an explosion), it is found to lie somewhat to the left of, and below, the middle of the main sequence in the H-R diagram. Astronomers have found another type of star which occupies this same region of the diagram and which may be related to the novae. These stars lie at the centres of slowly expanding spheres of gas. Because the gas spheres look rather like planets through a telescope, they have been labelled *planetary nebula*. It is thought that the central stars of planetary nebulae have exploded in times past and we are now seeing the result. If this interpretation is correct, the explosions must be quite infrequent: there are no records of planetary nebulae ever being observed to form.

No supernova has yet been seen before its explosion; so we really have no idea what sort of star produces it. At their brightest, supernovae are some two hundred million times brighter than the Sun, and are much brighter than the brightest nova. Nevertheless, as Fig. 21 shows, they fade away in a rather similar fashion. They are also much rarer in occurrence than novae: two or three dozen novae outbursts occur each year in our Galaxy, there may be only one supernova in two or three hundred years. It is evident that a supernova explosion is a much larger scale affair than a nova. A nova simply blows off its surface layers, leaving an easily recognizable star behind; a supernova may blow itself completely to pieces.

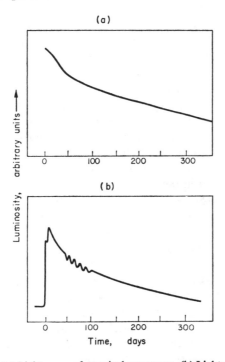

FIG. 21. (a) Light curve of a typical supernova. (b) Light curve of a typical nova. (It must be remembered that, in fact, the supernova is much more luminous than the nova.)

The last supernova to be seen in our own Galaxy appeared in 1604. Our knowledge of supernova explosions is therefore mainly gathered from observations of other galaxies. Nowadays, there is an international search programme to try and ensure that as many supernovae as possible are caught. Fortunately, supernovae are easy to recognize — they are so bright that, at maximum, they may contribute ten per cent of the total light emitted by a galaxy.

We can, alternatively, look for remnants of supernovae within our own Galaxy and examine their properties. The easiest way of detecting such remnants is from the radio observations. A supernova explosion ejects a cloud of material into space: like a nova explosion but with much more material and much faster (speeds of 2000–3000 miles per second are common). As this cloud moves outwards it radiates a good deal of energy — an appreciable fraction being in the form of radio waves. The 1604 supernova is rather dull visually: only a few wisps of gas can be seen. At radio wavelengths, however, it is a striking feature of the sky. On the other hand, the Crab nebula (so called because of its telescopic appearance) is much more interesting. It is the remnants of a supernova which exploded in 1054, and is prominent both visually and in the radio region.

It will obviously be of importance in stellar evolution to assign these exploding stars to their correct Population type. The effects of the explosion so complicate the spectra of these stars that it is virtually impossible to determine their chemical compositions. Our best method of approach therefore is to study their distribution within galaxies: whether in the arms, or in the nucleus. A fair number of novae have been observed within our own Galaxy. Although they evidently occur in the spiral arms, there is a general tendency for them to be concentrated towards the galactic centre. This suggests that their properties may conform more with Population II characteristics than Population I. This suggestion is confirmed by the fact that novae have been observed to flare up in two globular clusters.

Very little can be said with certainty about the distribution of supernovae. One major complication is that there are at least two types of supernova. The Type I supernovae probably belong to Population II. Spectra taken during the explosion of these stars suggest that they contain very little hydrogen. The explosion ejects about one-tenth of a solar mass into space. The Type II supernovae probably belong to Population I. They contain a considerable amount of hydrogen, and lose a quantity of material equivalent to several solar masses when they explode. The Type I supernovae seem appreciably brighter at maximum than the Type II.

It is important that some, perhaps all, novae are members of close double stars. One practical result is that it becomes possible to estimate the masses of the stars involved. This helps track down the stage in the evolution of a star when it is likely to explode. The theoretical implications are even more significant, for it appears that interaction between stars can greatly affect their development. Since an appreciable fraction of all stars have companions, we will need to keep this possibility in mind as we discuss stellar evolution.

Spectroscopic Peculiarities

Besides the large group of stars that vary in brightness, there are stars which are considered peculiar because they differ from normal stars in some other way — usually spectroscopic. For example, several types of star have spectra which vary with time. Thus, there is a group known as the peculiar A stars (usually abbreviated to Ap stars). These have spectra which are basically similar to normal A stars, but certain of the spectral lines — due to chromium, strontium and silicon, for example — are much stronger than they should be. In some Ap stars the lines vary regularly in intensity — sometimes very dense and black and sometimes appreciably fainter. These stars — known as spectrum variables — are of particular interest because they have been shown during the past few years to have very strong magnetic fields (several thousand times larger than the Earth's) which vary in intensity at the same rate as the spectral lines. Here is a stellar characteristic that we have previously overlooked. Are magnetic fields of any effect in stellar evolution? We must consider this question in a later chapter.

Another group which show time-variable spectra are the Be stars ('B' is the spectral type of the star on the Harvard classification and 'e' stands for emission). Apart from the normal black absorption lines of all B stars, the spectra of Be stars also contain a certain number of bright emission lines which vary irregularly with time. A close study of these stars has shown that they are actually spinning round very rapidly on their axes. (The speed at their equators is some 300 miles per second, whereas the equator of the Earth is only moving at a ¼ mile per second.) As a result, the gravitational attraction of one of these stars for material at its equator is barely equal to the centrifugal force outwards. Material is therefore continually lost into space. For a time it forms a ring round the central star, then it dissipates. Whilst forming, it produces emission lines in the stellar spectrum. These wax as the material congregates together, and wane as it expands away. Here then is another stellar characteristic which we have not yet taken into account. Can fast rotation effect stellar evolution? Again this is a question which we must examine in more detail later.

Besides being peculiar in the sense that they change with time, stellar spectra can also be peculiar in that they indicate unusual chemical compositions. What we are interested in here is not an overall change in chemical composition, such as distinguishes the Population I stars from the Population II, but abnormal abundances of a few of the elements relative to the remainder. The Ap stars, as we have seen, are peculiar in both senses — they have excessively strong lines of certain elements (which is usually taken to mean that these elements are over-abundant) and the lines may vary with time. It is generally assumed that the odd abundances in Ap stars are created by their magnetic fields, but there is no agreement yet as to the mechanism involved.

There is another, subtler, way in which the chemical composition of a star can be peculiar. Most elements in nature consist of mixtures of isotopes.* If we compare the relative proportions of the different isotopes present on the Earth, the Sun and the stars, we find that they are, in general, remarkably similar. But occa-

---

* Isotopes are atoms all of which have the same chemical properties, but which differ slightly in weight.

sional remarkable discrepancies have been observed. For example, helium in nature usually consists mainly of the heavier isotope He⁴, with a slight admixture of the lighter isotope He³.† One Ap star has been found, however, in which He³ is far more abundant than He⁴.

Another element in which large isotopic variations have been found is carbon. The lighter isotope here is C¹² and the heavier C¹³. In most stars (including the Sun) there are about ninety of the lighter carbon atoms present for every one of the heavier. Certain peculiar red giant stars have been found to have an excessive amount of carbon in their atmospheres. When this carbon is examined in detail it is found to consist of one of the heavier atoms to every three or four of the lighter.

In recent years, as methods of observation have become increasingly more accurate, many different types of spectral peculiarity have come to light. Unfortunately, they have proved very difficult to interpret. For example, there seldom seems to be any close link between the spectral peculiarity of a star and its position in the H-R diagram. The point to be remembered is that these abundance differences are seen on the stellar surface, whereas the evolution of a star depends on the happenings in its interior. Unless, therefore, the spectral peculiarities extend throughout the whole of the star, they will be superficial — in all senses of the word — so far as the study of stellar evolution is concerned. What astronomers must try and do eventually is to disentangle the spectral peculiarities which represent a fundamental alteration in a star from those which are purely atmospheric. This is likely to be a long task, depending heavily on theory: it is only in the first stages of development at present.

Summary

In this chapter we have been considering the various main groups and families of stars observed within our Galaxy. We

† He is the chemical symbol for helium. The superscript represents the number of particles in the nucleus of the atom.

have also noticed the appearance of occasional unusual stars which do not fit into the major groupings. In working out a theory of stellar evolution, our first task is to explain these major groups. We can use the peculiar stars, however, to guide our thinking, and to fill in gaps in the general picture.

CHAPTER 3

The Structure of Stars

ALTHOUGH stars are extremely diverse, nevertheless there are certain structural features which they share in common. In the same way plants may take the most various forms, yet they can generally be said to possess roots, leaves, etc. The most obvious feature of a star is that it emits energy. This, indeed, is the criterion we use to divide stars from planets. It is evident from the geological history of the Earth that the Sun has been emitting much the same amount of radiation for several hundreds of millions of years into the past. We can assume, therefore, that stars in general are capable of producing energy over prolonged periods of time. We are thus forced to suppose that stars contain huge reserves of energy. The processes producing this energy must, moreover, be able to operate without disturbing the overall equilibrium of the star.

The State of Stellar Matter

Is the material in a star solid, liquid or gaseous? It can hardly be solid. The surface temperature of the Sun is high enough to melt any known material, and it is certain that the temperatures within the Sun are even higher. The possibility that stars are liquid seems at first more promising. A rough theoretical estimate shows that the material at the centre of the Sun should be considerably denser than water. Gases do not usually have densities as high as this. But we must remember that the centre of the Sun is very hot (it is at a temperature of about ten million degrees). Now the temperature of a body can usually be taken as a measure of the speed with which the atoms in it are moving. And the atoms as they move about, strike against each other frequently. The number of collisions and their violence will increase rapidly with temperature. At a temperature of a few million degrees, the atoms will be moving very fast indeed. The violence of the collisions is great enough to disrupt most atoms present, and break them down into sub-atomic particles. Because the electrons in an atom cannot approach closer than a certain distance from the central nucleus, the atom, as a whole, takes up a good deal more space than the particles which form it. It is like a large building which may take up a considerable amount of space only a small part of which is actually occupied by the walls and the floors. We can pursue this analogy further. If the building is demolished, the volume of space occupied by the rubble is much smaller than the volume of the original building. So it is with atoms: once they are broken down into electrons and nuclei, they can, in effect, be jammed much closer together without undue squashing. A substance becomes a solid or a liquid when its constituent particles are jostled close enough together. Gases, on the other hand, represent the state where the particles have plenty of freedom for movement. The effect of high temperatures within a star is therefore, roughly speaking, to turn what should be a solid or liquid into a gas. We have been talking here mainly of the conditions at the centre of a star. We find on a detailed examination, however, that at points farther from the centre, although the temperature is lower, so is the density. The stellar material at these points is therefore still gaseous. We can say in summary that a star is basically a sphere of gas.

At any point within a normal star the material must be more or less in equilibrium (not moving very rapidly). If this were not true, the star would be unstable — we would see great changes over short periods of time. In equilibrium, the weight of the material pressing downwards at a given point is balanced by the pressure in the material below pushing upwards. (In a similar way, a man standing on a floor is in equilibrium because his weight, which is pulling him downwards, is balanced by the strength of the floor boards which are pushing him upwards.) Obviously the further into the star we go, the greater the weight of the material overhead, and so the greater the gas pressure needed to sustain it. It is one of the characteristics of gases — as distinct from liquids and solids — that they can be very readily compressed. As a result the density of stellar material increases very markedly towards the centre of a star: as has been pointed out in the previous paragraph. On the other hand, the density falls off very rapidly towards the surface of a star — the solar atmosphere has a much lower density than the terrestrial atmosphere.

The Transfer of Energy in a Star

As we shall see later in this chapter, the energy that a star emits is produced in a small region round its centre. We must therefore consider how the energy passes from the centre of a star to its surface. There are three possibilities — by radiation, by convection, or by conduction.

If the energy is transported by radiation, we can think of the stellar material as being fixed in position. The radiation trickles out through the material like water percolating through a sponge. The journey from the centre of a star to its surface is frequently interrupted: as the radiation passes outwards, it is continually being absorbed or scattered by atoms and sub-atomic particles. This ability of stellar material to stop radiation is called its opacity. Even by terrestrial standards stellar material is quite opaque. On the average, radiation in the stellar interior travels for about an inch before being absorbed — which corresponds to an opacity like that of a muddy pond. This high opacity produces two results. In the first place, the radiation is hindered in its attempt to leave the star. It takes many years to leak out from the centre to the surface. Secondly, the radiation gradually loses energy as it filters out from the centre. The amount of energy possessed by a given quantity of radiation depends strictly on its frequency: the higher the frequency, the greater the energy. At the centre of a star most of the energy is in the form of high-frequency X-rays. By the time the radiation has worked its way out to the stellar surface, it has been converted to much lower frequencies, appearing mainly as light.

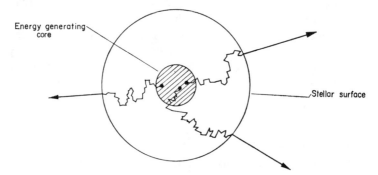

Energy generating
core

Stellar surface

FIG. 22. Passage of radiation through a star.

Energy can also be transported through a star by convection. It is this process which occurs when water boils in a kettle. The water at the bottom of the kettle is warmed by heat from the stove, and therefore expands. The expansion makes it lighter than its surroundings and so it rises upwards. It is immediately replaced by colder water from above which, being heavier, streams downwards to fill the vacancy. This colder water is heated in turn and rises. After a while, a continuous cycle of ascending and

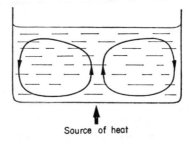

Source of heat

FIG. 23. Convection currents in boiling water.

descending water is established. This represents a convection current transporting energy from the stove throughout the water. The same sort of mechanism can occur in stars. The source of heat is now the centre of the star, where energy is being generated. The stellar material plays the part of the water in the kettle.

Transfer of energy by convection obviously requires that the stellar material, itself, must move. In this it differs fundamentally from radiative transfer. The motion is, however, slow by astronomical standards — less than 50 miles per hour — so it does not upset the equilibrium of the star.

It is important to decide how the energy transfer depends on the conditions prevailing within a star. In general, energy is transported from point to point by radiation (even when most of the energy is being carried by convection currents, a small amount is transported by radiation). However, should the stellar material for some reason be very opaque, transport of energy by radiation becomes most inefficient. The heat from the centre is dammed up by the opaque material; so, in order to reach the surface, it circumvents the obstacle by setting up convection currents. In going from the centre of a star to its surface, there may be certain regions where the energy is transported by convection and others where it is transported by radiation.

So far we have not mentioned the third method of transporting energy — conduction. (This is the process by which a poker with one end in the fire becomes hot at the other end.) It has been left to one side because it is only important in astronomy under special circumstances. The reason is simply that the best conductors in nature tend to be solids (metals, such as copper, for example). But we have seen that stars are made up of gaseous material. It is possible for material in some stars, however, to become much more compressed than usual. Under these conditions, even the sub-atomic particles are pressed for space. The stellar material now begins to act much more like a solid than a gas, despite its high temperature. It may, in particular, turn into an excellent conductor. When, in the future, we have to deal with very dense stars, we must therefore allow for the transmission of energy by conduction.

Energy Production in Stars

Until just before the Second World War, the origin of stellar energy remained a mystery. Now there is general agreement (although some details are still in dispute): stars derive their energy from nuclear reactions.

As we have remarked, the centre of a star is very hot. Most atoms in the central regions are completely broken down into sub-atomic particles, and these particles fly about at high speeds frequently colliding with each other. The electrons are not affected by these collisions, but the atomic nuclei — as has been shown by laboratory experiments — can interact under these conditions. Different nuclei do not interact with equal ease. As a very rough rule of thumb we can say that the lighter the nucleus, the more rapidly it interacts with other nuclei. This is partly because the lighter nuclei move faster. It is also partly due, however, to the fact that all nuclei have a positive electric charge. The heavier nuclei have larger electrical charges than the lighter nuclei; when they approach each other they therefore experience a greater mutual repulsion. This keeps them apart and prevents interaction.

Generally speaking, a star maintains its central temperature at just a high enough level for the lightest element present to interact. The lightest element present at the centre of the Sun is hydrogen. The hydrogen nuclei (protons) are colliding more frequently and with greater energy than any of the others and they are therefore interacting. (The rate of interaction is actually fantastically slow by normal standards: a proton will have to wait, on the average, more than a thousand million years before it interacts with another proton. Even so, the Sun contains such a vast amount of hydrogen that these intrinsically rare interactions produce over a hundred billion billion* horsepower.)

* A billion here means a million millions.

When one proton interacts with another, it first produces the heavy isotope of hydrogen, which is known as deuterium. The deuterium then interacts very quickly with another proton to produce a nucleus of the light isotope of helium (He³). The final stage of the build-up can follow more than one path. One possibility is that two of the He³ nuclei interact to produce a normal helium nucleus (He⁴) plus two protons. Whatever the ultimate step, the overall result of the interactions is to produce one helium nucleus from four protons. But four protons weigh very slightly more than one helium nucleus, so the interactions have caused a certain amount of mass to disappear. As Einstein showed early in the present century, if mass disappears, energy must appear to take its place. This is the source of the energy in an atomic bomb and it is equally the source of the energy in a star. The energy can appear in various forms. Some appears as very high frequency radiation. Some speeds up the interacting nuclei and makes them move about more rapidly. Some is expended in creating new sub-atomic particles, which may, or may not, produce further reactions of their own. The whole process from the initial interaction of two protons to the final production of a normal helium nucleus is called the *proton-proton* chain.

In stars which are appreciably hotter than the Sun, the protons combine to form helium nuclei in a different way. They now use the carbon nuclei present as a catalyst.* A proton combines with a normal carbon nucleus to form a heavy isotope of carbon. This interacts with another proton to form the light isotope of nitrogen. The addition of a third proton forms the heavy isotope of nitrogen. But when a fourth proton is added to the heavy nitrogen, instead of building up further into oxygen, it breaks down into a helium nucleus plus the original carbon nucleus. The carbon nucleus can now be used again. The overall result of this process is again to convert four protons to one helium nucleus, and energy results as in the proton-proton chain. This new series of nuclear reactions is called the *carbon cycle,* or, more accurately, the CNO cycle,

---

* A catalyst, in its widest sense, is a substance which takes part in a reaction, but which reappears unchanged at the end.

since nitrogen (N) and oxygen (O) participate along with carbon (C).

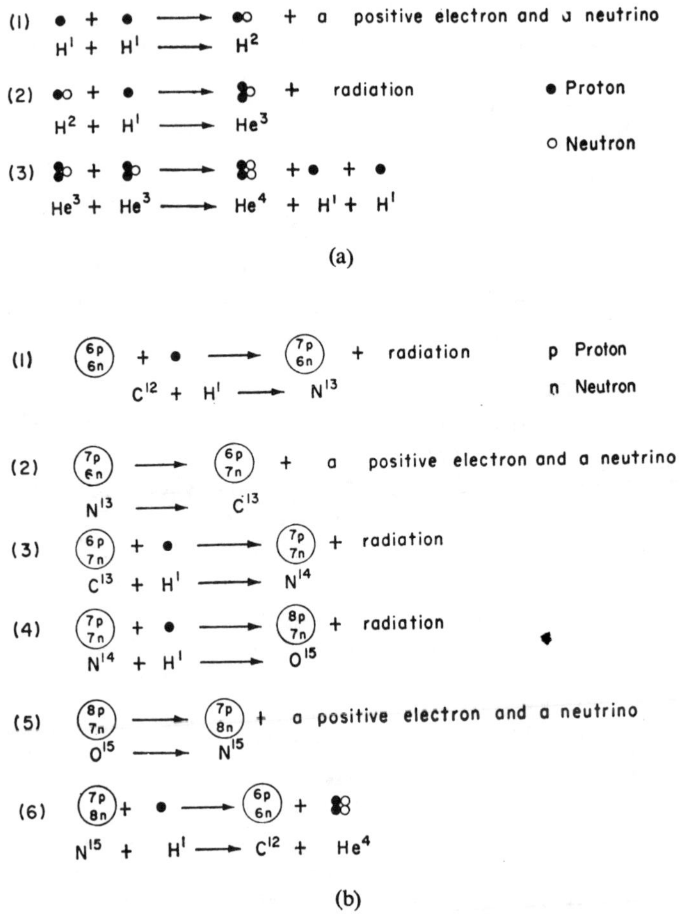

(a)

(b)

FIG. 24. (a) The proton-proton (p-p) chain. (b) The carbon (CNO) cycle. The sequences shown here are the simplest forms. The reactions involved may change in detail as conditions in stellar interiors vary.

So far we have been considering a reacting gas mixture with hydrogen as its main constituent. What reactions are possible if there is no hydrogen present? The next lightest element after hydrogen is helium. If the temperature is raised sufficiently — from ten million degrees to a hundred million degrees — the helium nuclei will begin to interact, forming beryllium. Here, however, we run into difficulty. The beryllium produced is highly unstable and breaks down immediately into the two original helium nuclei. In order to build helium up into a heavier element, it is necessary to have three helium nuclei hitting each other at more or less the same instant. A triple collision will fuse the three nuclei together into a stable carbon nucleus. Since one carbon nucleus weighs less than three helium nuclei, energy will be produced as before. Triple collisions are rare under normal circumstances, so helium will only fuse into carbon at very high densities and temperatures. This helium-burning process is called the *triple-alpha* reaction (because physicists usually refer to a helium nucleus as an alpha particle).

At a temperature little higher than that required for the triple-alpha reaction, other nuclear reactions can also occur. Helium nuclei can react with carbon to produce oxygen. More helium nuclei can react with this oxygen to form neon, and so on to heavier elements. The carbon formed from the triple-alpha reaction is the normal isotope containing twelve sub-atomic particles. This is also the carbon isotope which initiates the carbon cycle when hydrogen is burnt. But the carbon cycle, itself, changes some of this carbon to the heavier isotope with thirteen particles. (We have seen in the previous chapter that this heavier isotope has actually been observed in stars.) The heavier isotope is also capable of reacting with a helium nucleus at high temperatures. Oxygen is produced again, but now it is accompanied by a neutron.* Because it bears no electrical charge, the neutron is not repelled by nuclei; it can enter them very easily and interact. This means that neutrons are very efficient in inducing nuclear transformations. For this reason, as soon as the temperature rises

* An electrically neutral particle with the same mass as a proton.

high enough for helium to react with carbon, a whole string of nuclear reactions become possible.

The picture now becomes highly confused. We will take up the details in a later chapter. For the moment, we will only consider the ultimate end of all these reactions. The transformation of hydrogen to helium produces energy. So does the build-up of helium into carbon; but the amount of energy produced by the latter reaction is smaller. Succeeding reactions continue to produce energy, but in smaller and smaller quantities as the nuclei grow heavier. Eventually the build-up reaches iron (with a nucleus containing 56 sub-atomic particles). At this point the nuclear reactions cease to produce any energy at all: to create a nucleus of higher weight than iron actually requires that energy be put into the process. But we have said that stellar energy derives from nuclear reactions. What happens then to a star that has burnt all the hydrogen at its centre to iron? Where is its source of energy? There is, in fact, another equally important and related question which we have hitherto ignored: how does a star become hot enough for hydrogen to burn in the first place? The answer to both these questions — one concerned with the last stages of a star's evolution, the other with its first stages — will be discussed in later chapters. In both cases we will find that the star must undergo rapid alterations in its structure. The assumption of equilibrium which has been implicit throughout the present chapter then breaks down.

Stellar Models

We have so far been discussing generally the factors which are capable of affecting stellar structure. Supposing we want now to calculate a model of a specific star, how do we set about it? As we saw at the end of Chapter 1, we must start by assuming a mass, chemical composition and age for the star. We then use information, acquired in the laboratory, concerning the nature of the physical processes involved (the factors we have been discussing in this chapter). Next we work out how a star with the assumed properties would appear to observation. Finally we compare the theoretical luminosity, surface temperature and radius with the observed values. Supposing that the theory and the observations agree, what precisely have we derived? Firstly, of course, we have derived a mass, chemical composition and age for the star. But we have also derived the structure of the star: we know where in the star energy is being transported by radiation, where by convection; we know what nuclear reactions are occurring at its centre; we know how the density, pressure and temperature vary from point to point. It is a basic theoretical tenet that the structure thus derived is unique — no other combination of mass and chemical composition will produce precisely the same structure. Unfortunately, this does not mean that any given star can immediately be matched up with a unique stellar model. It is possible to calculate two or more alternative models of widely differing structures (and therefore widely differing masses and chemical compositions) which nevertheless have the same luminosities and surface temperatures. To observation, therefore, they will appear the same, unless we can draw on some independent data to clarify the position.

Most stellar masses probably remain constant over long periods (that is, the average star does not gain or lose much material during most of its lifetime). The chemical composition, on the contrary, is constantly changing due to nuclear reactions in the interior. In constructing a stellar model it is therefore not enough just to assume an overall chemical composition: the variation from point to point through the star must also be specified. The simplest case is a star which has only just been born. We may suppose that the birth of a star corresponds roughly to the instant when nuclear reactions start at its centre. It also seems reasonable to suppose that a star at the beginning of its career is homogeneous throughout. In other words, a zero-age star has a uniform chemical composition from its centre to its surface. As we have seen, our knowledge of chemical compositions is derived entirely from observations of stellar surfaces. For a zero-age star this gives us immediately the composition of the whole star. But what if we are dealing (as we usually are) with stars of finite age? So long as the stars appear to be normal, we may tentatively assume that the surface has not undergone much change since the star was formed. The present observed composition of the surface is then the original composition of the whole star.

A zero-age star with a uniform chemical composition obviously provides the simplest possible stellar model. To calculate such a model we set up a series of equations which describe what is happening at any point within the star. (For example, we can write down an equation to show how the rate of energy production depends on the temperature and density at any given distance from the centre of the star.) We then take all the equations and solve them simultaneously to give us our stellar model.

In the course of the computation we have to keep in mind all the various possibilities for the structure of our model. We must ask, for example, how the energy is being transported at each point; depending on the answer, quite different stellar models will result. A star where the energy is transported mainly by convection will have a higher surface temperature than one of the same mass whose energy transfer is mainly by radiation. If the star contains plenty of hydrogen, and is using this as its chief fuel, we must also decide which nuclear reaction is occurring. Is it the proton-proton chain or the carbon cycle? Again, depending on the answer, the resultant stellar models will be quite different.

Supposing, however, that we have selected the proper conditions for our star and end up with a satisfactory stellar model. What are its basic characteristics? The most obvious is that the pressure, temperature and density all decrease rapidly and continuously from the centre to the surface. A rather more surprising result is that the energy is almost entirely produced within a quite small region round the centre of the star — within the first 25 per cent of the radius for the Sun; within the first 5 per cent for a star with ten times the Sun's mass ($10 M_\odot$). This concentration towards the centre is due to the very strong dependence of the nuclear reaction rate on the temperature: below a certain temperature there is virtually no interaction. Similarly, though to a less extent, the material in a star is concentrated towards the centre. For the Sun, over 90 per cent of the material lies within half a radius from the centre; for a star of $10M_o$ the figure is much the same. Perhaps one of the oddest results is that the less massive stars are denser at their centres than the more massive ones. Thus the material at the centre of the Sun has a density over a hundred times that of water, but the material at the centre of a star of $10M_o$ is only ten times as dense as water.

So far we have been considering the results for computations of zero-age stellar models. This is only the beginning of our study of stellar evolution. Few of the stars in the sky have just been born — most are thousands of millions of years old. But we can use our zero-age models as stepping stones to the older stars. We have seen that nuclear reactions change the chemical composition in a star's interior. These changes alter in turn the luminosity and surface temperature of the star, so that it differs from the zero-age star. (This is equivalent to saying that evolution changes the position of a star in the Hertzprung-Russell diagram.) On the other hand, the chemical composition of the stellar surface remains much the same. We can therefore adopt the following approach. We determine the chemical composition of the stellar surface and calculate a zero-age stellar model which has this composition throughout. Its surface temperature and luminosity will not agree with those of the observed star. We next look at the distribution of the density and temperature in this first model. We can now estimate how fast hydrogen is being converted to helium at any point in the star. From this we can determine how the composition is changing throughout the star. This knowledge enables us to calculate a new stellar model with a chemical composition that changes from the centre to the surface Our second model has a different run of density and temperature from the centre to the surface from the first. We use these new values to help us estimate once again the rate at which hydrogen is being consumed at any point in the star. This leads to the construction of yet a third model with different temperatures and densities, and so on. Each new model represents a step forward in time: how great a step we can take will depend on the particular star, but a reasonable figure for a star like the Sun would be five hundred million years per step. We continue this sequence of *inhomogeneous* models, until we find one which gives the same luminosity and surface temperatures as the star we are considering. This then constitutes our final stellar model for that star. Note that this gives us not only the mass and chemical composition of the star, but also its age.

There is one further complication which has yet to be mentioned. Before we can work out how changes in chemical composition will affect the future of a star, we must know whether the material in a star is stationary, or whether it moves about. This usually reduces to a question of the energy transfer within a star — whether it is by convection or radiation. If the energy transfer is by means of radiation (or, for that matter, by conduction) the material remains where it is. A change in chemical composition will only affect the point where it occurs. If, on the other hand, the energy transfer is by convection, the material gradually moves and any change in chemical composition will be distributed over the whole region where convection is active. In other words, there is no mixing of material in a radiative region. Hence nuclear reactions produce inhomogeneities in the chemical composition. In a convective region mixing occurs. Nuclear reactions here result in a homogeneous zone with a slightly altered chemical composition throughout. Since the future evolution of a star depends strongly on the chemical composition at each point within it, the way in which energy is transferred has a considerable effect on its development. As a star evolves, it may well, of course, change its method of transporting energy. It is necessary to keep a sharp eye open for such changes when computing a series of stellar models. Should a star be transporting energy by convection throughout a major part of its volume, one of our original assumptions is likely to break down. Material from the stellar interior, which has been processed by nuclear reactions, may be brought up to the surface by the convection currents and mixed in with the original surface material. This would, of course, affect the chemical composition we assume for the initial zero-age model.

Stellar Models and Electronic Computers

Extensive computations of accurate stellar models have only been made during the last few decades. For, as was pointed out at the end of Chapter 1, the calculation of stellar models requires the use of large electronic computers. The reason is that the equations which must be solved to produce a stellar model have no simple solutions (none, at least, that correspond to very realistic stars). The numerical computations involved are therefore very heavy. On an electronic computer the calculations may now take only a minor amount of time — a few hours, perhaps. Most of the effort is in putting the calculations into a form that the computer will accept. Even so, there are stages in the evolution of a star — where it is changing its internal structure rapidly — that are exceedingly difficult to chart. Attempts to model these stages require the resources of the largest computers currently available, and the end results remain approximate.

CHAPTER 4

The Birth of Stars

The Stars and Interstellar Material

It has been generally held from the early days of work on stellar evolution that stars must in some way originate from the diffuse interstellar medium. It is, of course, rather difficult to see where else they could have come from; but, apart from this rather negative approach, there are sound observational reasons for accepting the hypothesis. Consider the following argument. The life span of a star depends basically on two factors: the rate at which it emits energy per unit time and the total amount of energy available for emission. These two together will obviously give the total possible lifetime of the star. The first factor is simply the luminosity of the star. The second depends on the quantity of material available for nuclear transmutation which is, in turn, proportional to the mass of the star. Thus a comparison of the mass and luminosity of a star will give an estimate of its life span. (The estimate is only approximate because the luminosity of a star does not remain constant from birth to death, as we are assuming here.) If we make such a comparison for the Sun, we find (as we might expect) that the Sun's reserves of energy are considerably more than are necessary to account for the present age of the solar system. Indeed, the Sun's stock of nuclear fuel should last it for some ten thousand million years altogether. If we turn our attention now to very bright stars, we find quite a different situation. Such stars pour their energy into space at a prodigious rate — perhaps a million times more per second than the Sun. As they are also more massive, they have a greater reserve of nuclear fuel than the Sun. Nevertheless, this is insufficient

66

to counterbalance their loss of energy, and their life is much shortened — they may last for only a few million years (which is a negligible period by astronomical standards.) The very bright stars in our Galaxy are therefore also very young and short-lived.

Since the bright stars must have been born quite recently, we might hope, by examining them, to find out something about the events attending their birth. In fact, all the observations demonstrate that these young stars are either embedded in clouds of dust and gas or, at least, have such clouds nearby.

We can fit this in very neatly with the birth of stars from the interstellar medium. A star is born in the middle of a cloud of gas and dust. Then the cloud gradually disperses and the star goes on alone. This latter stage, when the gas and dust move away from the star, may, indeed, be observed. We can even explain why such a dispersion should occur. The radiation pouring out from the bright stars exerts a pressure on the material surrounding them (just as the energy emitted by the Sun exerts a pressure — which can be measured — on artificial Earth satellites). This radiation pressure pushes the interstellar material away from the star, leaving it eventually isolated in space.

Motions in the Interstellar Medium

We will accept then that the observational evidence indicates a connection between young stars and the interstellar material. If this is due to the one being derived from the other, how does the transmutation take place? The essential clue is contained in the very obvious fact that the interstellar medium is diffuse, whereas stars are very compact. We can go from the first state to the second by condensation. What we need therefore is a force that will draw diffuse material together. Such a force undoubtedly exists in gravitation — the inherent tendency of matter to attract other matter. In fact, at first sight it might seem that gravitation solves the problem too efficiently. We might reasonably ask why, if all matter is mutually attractive, the interstellar medium did not condense entirely long ago. However, an examination of the solar system with this question in mind soon provides an answer. The planets are gravitationally attracted to the Sun, but they do not crash into it; the reason is simply that they are in motion, and the centrifugal force outwards balances the gravitational pull inwards. A simple analogy would be a stone tied to the end of a piece of string and whirled round rapidly. The string is in tension due to the circular motion of the stone (Fig. 25). (We can deduce that if the circular motion of the planets was suddenly cancelled, they would immediately fall into the Sun because there would be nothing to oppose the gravitational attraction.) Similarly, if the interstellar medium is in motion, this will provide a good reason for its continued existence. But we have, in fact, already observed that

Centre

FIG. 25. The tendency for a body moving in a circle to fly off at a tangent can only be prevented by a force inwards.

our Galaxy is rotating, so the interstellar material is in motion round the centre of the Galaxy. Besides this large-scale motion the interstellar clouds also have small random motions both internally and relative to each other.

It is worth examining these random motions in more detail, since they can tell us something of the conditions in interstellar space. Interstellar material, like stars, can produce spectral lines. Like stellar lines, these may be bright or dark. Both types are produced by the gas — the dust present has very little effect spectroscopically. If the gas is hot, bright lines are emitted. For example, gas which is being pushed away from the vicinity of young stars is heated in the process and emits bright spectral lines. On the other hand, suppose we examine the spectrum of a distant star. We will find, besides the dark absorption lines formed in the stellar atmosphere, other lines formed by absorption in the interstellar gas. If we inspect these interstellar lines we find that they are frequently not single, but split up into several components. This is due to the nature of the interstellar medium; it is not evenly distributed throughout the spiral arms, but is concentrated into clouds. The random motions of the clouds produce absorption lines at slightly different wavelengths due to the Doppler effect.

A study of these interstellar lines enables us to deduce two important pieces of information about the clouds. In the first place, we can find the density in a cloud. An average sort of value

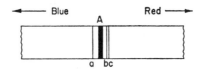

FIG. 26. Interstellar absorption lines. The line A is due to the star being seen through three interstellar gas clouds (a, b and c) each of which is moving with a different speed.

would be 10,000 atoms per cubic inch (as compared with only one per cubic inch between the clouds). Secondly, we can estimate the sizes of the clouds. They cover a considerable range, but an average cloud might be 30–40 light-years across. These figures allow us to deduce that an average cloud has a thousand times the mass of the Sun. Most of the clouds have low temperatures — say 200°C below freezing* — but those which are near bright stars may be considerably heated to temperatures of over 10,000°C. Thus the clouds contain a certain amount of energy in the form of heat. They also have a certain amount of energy due to aerodynamic motions within the clouds — rather like the winds blowing in the Earth's atmosphere.

* It is customary to define an absolute minimum temperature at about 273°C below freezing. Thus the clouds, though very cold, still retain a little heat.

The Effect of Heat Energy on Condensation

We must now consider the gravitational interaction in such a cloud. Every particle is attracting every other particle so there is an overall tendency for the cloud to condense to a smaller size. The total gravitational force acting will depend on the number of particles in the cloud and their distance apart. Or, in other words, the tendency a cloud has to contract depends on its mass and its density — the greater the mass, the greater the contraction force; the lower the density, the less the contraction force. Now the internal energy in a gas cloud opposes any tendency it has to contract. (The heat of the cloud is an indication that the atomic particles which form it are in motion. They resist any effect — such as contraction — which tries to control their motion.) The fate of an interstellar cloud therefore depends on which is greater: its gravitational energy or its internal energy. A simple calculation shows that for an average cloud these actually balance: the gravitational attraction trying to contract the cloud is equal to the internal energy trying to expand the cloud. Thus most of the gas clouds in our Galaxy are in equilibrium and are therefore not forming into stars at all. Notice, incidentally, that even if one of these clouds should for some reason condense, it would presumably form into a star one thousand times heavier than the Sun: which is many times more massive than any star found in our own Galaxy.

The Effect of Magnetic Energy on Condensation

The spiral arms of our Galaxy are threaded by a magnetic field. We can best understand its significance for our present problem by considering briefly the current research work aimed at reproducing stellar energy generation mechanisms in the laboratory. This is usually called *thermonuclear* research. The idea is simply to heat up a small volume of gas until it reaches a temperature similar to that at the centre of the Sun. At this point the nuclei should interact producing energy. Unfortunately, as soon as a small volume of gas is strongly heated it expands and disperses. It must therefore be contained somehow during the heating process. (The containment in stars is effected by the gravitational forces holding the material together. But gravitational forces only become powerful when there is a sufficient mass of material present. They are completely inadequate for the small amounts used in laboratory experiments.) If a material container is used — say, a metal box — the heat pumped into the gas is immediately communicated to the box and thence to its surroundings. It is impossible to heat the gas to any very high temperature by this method.

As we have seen, nuclei only begin to interact when the temperature 1 is risen high enough for all their electrons to be stripped away. The particles present in such a gas therefore all have electrical charges — either positive or negative. A hot, ionized gas of this type is called a *plasma*. As it consists of an equal number of positive and negative charges, the gas as a whole

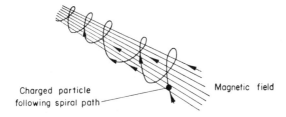

Charged particle
following spiral path

Magnetic field

FIG. 27. Motion of a charged particle in a magnetic field.

is electrically neutral. In studying plasmas — as distinct from ordinary gases — a significant new factor appears. The motions of charged particles can be altered by the presence of a magnetic field, whereas those of neutral particles cannot. If we project a charged particle at a magnetic field, we find that its motion across the field tends to be halted, and the particle is guided so that it moves in a spiral path preferentially along the field (Fig. 27). Suppose then we take a plasma and surround it by a magnetic field. The plasma will try to expand outwards, as before, but the magnetic field resists this motion of particles across it. The plasma is therefore held in a magnetic 'container' which will not dissipate the heat like a material container.

In the laboratory, magnetic fields can be used to stop a hot gas expanding. They can perform the same function on an astronomical scale: they are, for example, an important influence in determining the shapes of spiral arms. But a magnetic field can also have the opposite effect: it can stop a plasma from contracting. Now the interstellar gas clouds we have been discussing are appreciably ionized. This means that all their movements are affected by the galactic magnetic fields. Suppose then that an interstellar cloud did try to contract. It would fall in freely enough along the magnetic field, but perpendicular to the field it would soon be brought to a halt. The end of the process would be a disk of material not a spherical star. Even if we could force the material into the shape of a sphere, it would contain so much magnetic field that it would immediately blow apart again.

The Effect of Rotational Energy on Condensation

We will return to our previous example of swinging a stone at the end of a piece of string, and now perform a new experiment. As the stone is circling round, we suddenly shorten the length of the string. Immediately the stone begins to swing round faster. Similarly, if a cloud of material, which is rotating, slowly condenses and becomes smaller, it will also begin to spin faster. Interstellar clouds are spinning round very slowly. (They have picked up this rotation from the rotation of the Galaxy as a whole.) It is an easy enough matter to estimate how fast a star would rotate if it condensed directly from an interstellar gas cloud. Its equator would actually be moving round at over 150,000 miles per second — approaching the speed of light. This is obviously completely ridiculous. We find, in comparison, that the equatorial speed of the Sun is about one mile per second. At speeds greater than a few hundred miles per second, the gravitational attraction at the star's surface is less than the centrifugal force outwards, and the star dissipates itself into space. Thus we cannot condense stars directly from interstellar clouds, because they would be rotationally unstable.

Possible Methods of Condensation

We are thus faced with a considerable problem. Our former problem of explaining why the interstellar medium does not immediately condense under its gravitational attraction has certainly been solved, but we seem to have shown, instead, that condensation into stars is impossible. However, this result should not be thought too discouraging. From the amount of interstellar material left in our Galaxy, despite its age, we must assume that the formation of stars only occurs under special circumstances. Our task now is to identify these circumstances.

One possibility is that our description of the material in interstellar space has been incomplete. If some of the material is present as large rocks, rather than as dust, most of the difficulties disappear. In the first place, rocks would not be affected by the galactic magnetic fields. It is also possible to devise ways in which an assemblage of rocks could lose their excess of rotation during condensation. Moreover, rocks could build up anything — from the smallest planet to the largest star — depending on the quantity available. Unfortunately, this attractive scheme has two drawbacks. First, there is no reasonable explanation for the origin of the rocks themselves. They would, presumably, have to be formed from the interstellar dust, and all the current indications are that such a process is inherently improbable. Secondly, if there are large chunks of rock in space, they should produce some observable effects. For example, light from distant stars should be blotted out altogether. We actually observe that starlight is dimmed in passing through interstellar clouds, but not usually obliterated completely. We can understand this if the clouds contain dust only, but not if they contain any appreciable number of rocks.

Another possibility is that some extra force is acting at the birth of stars which compresses the interstellar material from outside. We have seen, for example, that a hot, bright star heats up the interstellar gas around it. This hot gas may expand and press on the cold gas further away, thereby compressing it. Or, again, some stars towards the end of their lives explode and eject material into space. If this ejected material encounters an interstellar cloud it will compress it. In both cases the theoretical calculations suggest that the degree of compression is too small to form stars. Nor is it obvious that these mechanisms surmount our previous difficulties with regard to magnetic fields and rotation. Nevertheless, there may be some supporting observational evidence. It has been found that small dark patches can sometimes be seen projected against the background of the hot, glowing gases round bright stars. These patches seem to be very dense globules of interstellar matter. They are small compared with normal interstellar clouds — often less than a light year across — and are about as massive as an average star. In other words, they appear to be ideal objects for producing stars, and their position suggests that they could have been formed by compression. On the other hand, compression is obviously not the only mechanism by which stars form, for it presupposes the prior existence of other stars to produce the compression. We must therefore seek for a more general cause.

We will start by re-examining the factors which prevent condensation of interstellar clouds. Consider first the presence of a magnetic field. This is only a problem if the gas cloud is ionized: an electrically neutral cloud can contract as though no magnetic field were present. Interstellar clouds are ionized because they are subjected to radiation from the stars: that part of the stellar radiation which has a high enough energy will split off electrons from the atoms in the clouds. If we could somehow shield the gas from the starlight, all would be well. Such shielding involves placing a thick layer of material all round the cloud. We can look at this in another way. Suppose we have a particularly large, dense, interstellar cloud. Then the outer layers will shield the inner, and the central region will be unionized because all the stellar radiation has been absorbed further out. Moreover, a cloud of this size will be dense enough and massive enough to condense under its own gravitational attraction despite its internal heat energy. But now we run up very acutely against one of our other difficulties. The kind of interstellar cloud we are now contemplating has the mass of a thousand Suns. How can this condense to form a single star?

Fragmentation of Gas Clouds

We must drop our assumption that stars are born individually. We must think of them, instead, as being born together in groups. If a gas cloud of $1000M_o$ condenses, we must suppose that it will produce perhaps a thousand stars like the Sun. They will appear in the following way. The large interstellar cloud contracts (or, at least, its central regions do). When it reaches a certain stage of condensation it becomes unstable. Pieces much smaller than the total size of the cloud are now capable of holding together and contracting under their own gravitation. At this point the cloud fragments. The resulting pieces continue to contract individually until they, too, become unstable. Further fragmentation then occurs. This process may recur several times. Contraction heats the gas; eventually the fragments will become so hot that they are recognizable as *proto-stars*. Fragmentation now ceases. This is because the material, which has previously been fairly transparent, starts to become opaque. As a result, the heat radiation instead of streaming out freely from the fragment, is held back. It therefore exerts a pressure on the material, opposing the contraction, and so prevents fragmentation.

This scheme of star formation fits in quite nicely with some of the observations. Many stars in our Galaxy do belong to clusters or associations. But what of those — such as our own Sun — which do not? We have seen that associations are unstable; after relatively short periods of time the stars can wander off individually into space. Theory also indicates that clusters are unstable, but, because the stars are grouped more closely together, they take considerably longer to disintegrate than associations. In fact, the rate at which any group of stars disintegrates depends on the number of stars in the group and their distance apart. A cluster with many stars close together will exist for a long time; a cluster with few members will disappear rapidly. This is well supported by the evidence. Theoretical estimates show that the massive globular clusters are, perhaps, ten thousand million years old. On the other hand, a medium-sized open cluster, like the Pleiades, may be only a hundred million years old.

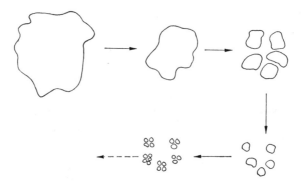

Fig. 28. The production of stars by fragmentation.

We would also imagine that, if stars are born together within a massive gas cloud, then the remnants of the original gas should still be visible during the early life of a cluster. This again we have seen to be true: young clusters can be distinguished from old by the presence of appreciable amounts of interstellar material. The material disperses gradually because it is blown away by the heating action of the newly born stars and the pressure of the radiation they emit.

If a large number of stars are born altogether, it is very likely that some pairs will be born at only a short distance apart, and will therefore exert a considerable mutual gravitational attraction. After the surplus interstellar material has blown away, such a pair will remain together, and we will see them as a double star. In some cases, there may be a third star in the near vicinity at the time of birth. This may remain attached to form a triple star. Multiple stars containing as many as six components have been found — Castor, one of the Gemini (the Heavenly Twins), is an example. We can think of these as multiple stars, or, if we like, as very small clusters. Obviously, this mechanism of forming stars in groups can account quite satisfactorily for the presence of a large number of double and multiple stars in the sky.

One interesting observation, still not completely understood, is that a few stars have been found moving at high speeds (100 miles per second or more) away from regions of active star formation. For example, three stars have been found moving away at high speed from the centre of the constellation Orion (which contains an association where many stars have recently been formed). These high-speed stars are young and hot, just like the ones which still remain in the association, and it seems only reasonable to suppose that they are former members that have been ejected. One possible explanation of these 'runaway' stars is that they originally formed one component of a double star. The other member of the pair blew up (we will see in a later chapter why this might occur) and its companion was thrown out into space. An explosion of this violence might be expected to produce other effects. It has been suggested, for example, that some faint, luminous filaments which can be seen round the main gas cloud in Orion (the Orion nebula) indicate places where waves of pressure from the exploding star ran through the surrounding gas.

Contraction of an Individual Star

Although stars may be formed in large groups, once a star has contracted far enough we can consider it as an isolated body. We can follow its future evolution individually and ignore how the other stars in the cluster are formed. (The only exceptions are close double stars, where the two components may come close enough together to interfere with their later development.)

The later stages in the birth of a star have been studied in much greater detail than the earlier stages. The proto-star, after it has undergone its last fragmentation, continues contracting under the influence of its own gravitational attraction. Its temperature continues to rise, for the gravitational energy lost by contraction reappears as heat energy. As the proto-star becomes opaque to the transmission of energy, the temperature rises much faster at the centre than at the surface. At this stage it is a cool, faint object which, if we plotted its position on a Hertzsprung-Russell diagram, would lie a long way off in the bottom right-hand corner.

The sequence of events is probably the following. Most of the material in the proto-star will be hydrogen. At low temperatures, hydrogen exists as molecules — that is as two hydrogen atoms linked together. As the temperature increases, the hydrogen molecules strike against each other with increasing frequency and violence. At about 1500°C, the collisional energy becomes sufficient to disrupt the molecules, and convert hydrogen into its atomic form. There is a lot of hydrogen present; therefore to disrupt it all, the proto-star must produce a large amount of energy quickly. This energy can only come from the contraction: to produce energy quickly, the star must contract quickly. In fact, the calculations show that the contraction at this stage turns into a rapid collapse inwards. When will this collapse stop? The obvious answer would seem to be — when all the hydrogen molecules present have been turned into atoms. A more detailed analysis shows, however, that this is untrue. The collapse, so to speak, generates too much energy. By the time all the hydrogen molecules have been pulled apart, the temperature has risen sufficiently for the hydrogen atoms themselves to begin, in turn, to split up into protons and electrons. Once this second process has started, the collapse will continue still further until it, too, has reached completion. But the collapse is still not over, even when the hydrogen has been completely ionized. As much as a quarter of the proto-star may consist of helium. This ionizes at a higher temperature than hydrogen. We can estimate that, when the collapse has gone far enough to break up the hydrogen atoms, it has also created a temperature high enough to initiate the ionization of helium. So the collapse continues until both the hydrogen and helium are completely ionized.

If we are considering the formation of a star like the Sun, the collapse will start when the proto-star has a radius about one hundred times the present distance of the Earth from the Sun. (This is well outside the limits of the planetary system as known at present.) The collapse stops when the proto-star has a radius about one-quarter of the Earth's distance from the Sun. (The proto-star then lies inside the orbit of Mercury, the innermost planet.) The entire collapse only takes a few hundred years: a very short period by astronomical standards. In the H-R diagram, the proto-star would suddenly jump from being faint and cool to being quite bright and appreciably hotter.

We must remember, however, that the proto-star is not isolated in space: it actually forms the dense core of a larger gas cloud. Hence, its evolution is a little more complicated than the preceding description implies. The proto-star forms first as a

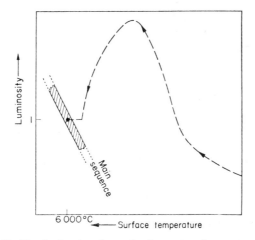

FIG. 29. Sketch showing the path of a contracting proto-star (of $1M_\odot$) in the Hertzsprung-Russell diagram.

small, dense core in the cloud. This collapses and, at the same time, attracts to itself additional material from the surrounding cloud. (The process is called 'accretion'.) Initially, radiation flows out from the core fairly easily, but as the central density increases, so the material in this region becomes more opaque. The central temperature therefore rises, slowing the contraction of the core until the break up of molecular hydrogen occurs, as previously described. All the time the core is contracting it is adding on further material from the cloud, so that its final mass is appreciably greater than the mass it had in its early years. For example, a core that would finally be massive enough to form the Sun would only weigh $\frac{1}{2}$M$_\odot$ after the first hundred thousand years of its formation.

For stars like the Sun, the opaqueness of the material forming the core is still quite high, even when the core has finally finished accreting. As a result, energy within the proto-star is transported by convection, rather than by radiation. Because convection is a very efficient method of transmitting energy the proto-star brightens considerably: the sudden kink upwards in its track in the H-R diagram is called its 'Hayashi track'. Continuing core contraction leads to a decrease in the opacity at the centre, so energy is transmitted there by radiation. Consequently, the star's track in the H-R diagram ceases to be mainly vertical (change in luminosity) and becomes more nearly horizontal (change in surface temperature).

It will be noted that, despite the various hiccups in the track of a solar-type star in the H-R diagram, the overall motion it follows is from right to left. This straightforward trend is even more obvious for more massive stars. For these, the central opacity decreases rapidly, and the Hayashi episode therefore disappears.

Observational Effects of Contraction

The early stages of star formation go quite rapidly once a core has formed: it is the last stages of contraction that take the longest. Even so, the total time for the formation of a solar-type star such as the Sun is only a few million years — a very short period in astronomical terms. Because we are dealing with stars that are changing their internal structures quickly, the computations are difficult to carry out and the results are approximate. However, one point seems certain: a star like the Sun will be fully convective at one stage in its formation. It will therefore be completely mixed, and so of uniform chemical composition when it arrives on the main sequence. If true, this has an immediate observational consequence.

When we were discussing nuclear fuels in the previous chapter, we first mentioned hydrogen-burning, then helium-burning, and finally jumped to the transmutations of carbon and heavier nuclei. But between helium and carbon there are three other elements — lithium, beryllium, and boron. To these may be added the heavy isotope of hydrogen — deuterium — which is always present in small amounts in terrestrial hydrogen. It is reasonable to ask where these elements fit into our scheme. The first point to emphasize is that these elements are very rare in nature; therefore, whatever their effect, it will be small. Laboratory experiments show that they all react more easily (that is, at a lower temperature) than hydrogen. They capture protons very readily and break up into helium nuclei. As a result, they can burn at relatively low temperatures — about a million degrees. Temperatures of this sort are reached in proto-stars during the final stages of their contraction. They can thus augment their gravitational energy by a certain amount of nuclear energy. The amount is small because of the scarcity of the elements. The effect on the track of the proto-star in the H-R diagram is therefore also small. However, the exact fate of these elements depends on the conditions within the proto-star. If the proto-star is completely convective for long enough all the material it contains will be mixed through the hot region at its centre, and then the elements we have been discussing will all be highly depleted, or even entirely destroyed. If, on the other hand, the proto-star is entirely radiative, the elements will only be depleted at the centre, and the surface layers will retain their primitive abundances. Hence a study of the content of deuterium, lithium, beryllium, and boron in a star may provide some clue to its history in the proto-star stage. We will make a detailed comparison when we discuss the Sun in Chapter 6.

If stars do, indeed, develop by contraction, as we have described, then amongst the multitude of stars in the sky some should be passing through this stage. What is the probability of finding one? The determining factor is the rate of stellar contraction. This is slowest when the proto-star is close to the main-sequence region in the H-R diagram; earlier, as we have seen, the contraction may be very rapid indeed. The actual length of time from the start of contraction to the final star depends mainly on the amount of contracting material present: the more massive a proto-star, the faster it contracts. The Sun takes several million years to contract. A star twice as massive as the Sun would take a

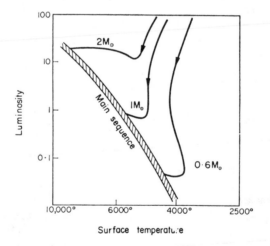

Fig. 30. Final contraction paths in the Hertzsprung-Russell diagram for proto-stars of different mass. The elbow in the track represents the point where the proto-star ceases to be fully convective. For masses less than $\frac{1}{4}M_\odot$, the proto-star remains completely convective right up to the main sequence, and the elbow disappears.

good deal less than a million years. On the other hand, a star with only half the mass of the Sun would take over a hundred million years. The mass also determines which regions of the H-R diagram are crossed by a proto-star during the course of its contraction. The more massive a star, the higher its path across the H-R diagram.

The probability of catching a star at a given stage in its evolutionary cycle depends on the duration of that stage. Since massive proto-stars contract very rapidly, the probability of spotting one is low. We stand a better chance of finding contracting stars of low mass since these take longer. We therefore look for fairly faint, red stars lying to the right of the main sequence in the H-R diagram. There is a major problem here, however. We will find later that stars may also move through this part of the diagram much later in their evolution. Any particular star in this region can, therefore, be either old or young, and it might be difficult to decide which from direct observation. We try to eliminate this difficulty by confining our search to young star clusters. We may then hope that any stars in the specified region of the H-R diagram are in the contraction stage. Indeed, if we do examine the H-R diagram of very young stars — the group around the Orion nebula, for example — we find a considerable number of faint stars above the main sequence.

Some of these are the T Tauri stars we have met before. As the name suggests, they have an (irregular) light variation. They are always associated with interstellar matter, which may be taken as further evidence of their youth. One fascinating point is that, unlike main-sequence stars, T Tauri stars are not deficient in lithium. If our picture of star formation is right, and the T Tauri stars are in the contraction stage, this implies that their central temperature has not yet risen high enough to burn lithium.

It may even be that the sudden jump in brightness of a contracting star, when it becomes fully convective, has been seen. Photographs of the regions round T Tauri stars frequently show what appear to be small knots in the interstellar medium. These are known as *Herbig-Haro objects* (after their joint discoverers). It was found, in 1954, that one of these objects in the Orion nebula apparently contained two stars which had certainly not been visible a few years before. We could suppose that they were originally very faint proto-stars which suddenly flared into visibility.

Since stars, when forming, are necessarily embedded in clouds of gas and dust, most young stars will be obscured from view. They can only be observed visually when much of the cloud has dispersed — that is, towards the end of their formation period. However, observations in the infrared can penetrate much further into clouds (because the wavelength of such radiation is greater than the size of many of the dust particles). Hence, forming stars can now be studied by infrared observations. It has been found, for example, that the Orion nebula, a well-known region of young stars, also contains strong infrared sources which may be proto-stars.

Rotation and Magnetic Fields

Earlier in this chapter we considered the forces which oppose the contraction of a star. We have since outlined a scheme of star formation which overcomes the opposition from the internal heat energy of the cloud and from its magnetic field. Now we must consider the third factor — the rotational energy. During the early stages of fragmentation excess rotational energy can be avoided rather simply. The energy which should have gone into making the fragments spin faster can be used instead to make them move round each other faster. In other words, spin energy can be converted into energy of motion. This only works, however, so long as fragmentation continues. As soon as the proto-star settles down to an individual career of contraction its spin rate starts to increase. By the time it reaches the vicinity of the main sequence, a star like the Sun should have an equatorial speed of several hundred miles per second. Since this contradicts the observations, there must be some further mechanism by which rotational energy can be lost.

We have seen that the magnetic field within our hypothetical interstellar cloud is ineffective because the material is electrically

neutral. We have also seen, however, that during the final contraction of a proto-star considerable heat is generated and the
material becomes ionized. At this point the magnetic field drops
its role of passive onlooker and once again intervenes. The
proto-star is now too greatly condensed for the magnetic field to
halt the contraction. Instead the field becomes anchored into the
star and is pulled inwards as the star contracts. The proto-star is
spinning round very rapidly: so fast, in fact, that it is shedding
material from its equator. As it contracts further, it leaves this
material, and the magnetic field it contains, behind. The magnetic
field is now anchored partly in the material and partly in the
proto-star. We can imagine lines of magnetic force stretching
b etween the two.

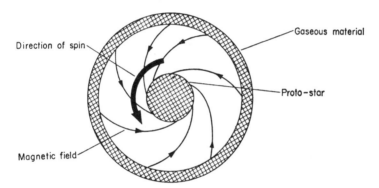

FIG. 31. The bicycle-wheel analogy.

We can picture what happens next most easily, perhaps, by a
simple analogy. Suppose the contracting star is represented by the
hub of a bicycle wheel. The material thrown off from its equator
is represented by the rim. The spokes of the wheel are the lines of
magnetic force buried in the star at one end and the material at
the other. To make our analogy complete, we suppose that the
spokes are made not of steel, but of elastic. If we rotate the hub
of such a wheel very rapidly, the first thing that happens is that
the elastic spokes are stretched out and wrapped round it. Their

tension then tugs at the rim until it, too, is set into motion. The rotation is being taken from the hub and transmitted to the rim via the elastic spokes. Similarly, we can argue that the rotation of a contracting star may be transferred to the material it leaves behind via the lines of magnetic force. This mechanism seems to be efficient enough to reduce the final rotational rates of stars down to reasonable values, as they near the main sequence: thus solving our last major problem of star formation.

CHAPTER 5

The Main Sequence

The End of Contraction

We have seen that proto-stars, during their contraction, move towards the main sequence. As they approach this region of the Hertzsprung-Russell diagram, their properties become increasingly similar to those of ordinary stars. The only difference is that their characteristics (brightness, surface temperature, etc.) are still changing slowly with time.

The brightness of a proto-star comes from its contraction, and this serves, too, to heat the star. By the time a contracting star reaches the neighbourhood of the main sequence, its central temperature has shot up to over a million degrees. Now, slowly, the hydrogen nuclei start to interact. It was in this way that the star, at a slightly earlier stage in its contraction, burnt up its stocks of deuterium. But ordinary hydrogen is much commoner and reacts more slowly than heavy hydrogen. So, whereas deuterium burning represents only a minor episode in a star's life, hydrogen burning takes up a major part of its life cycle.

The energy derived from the conversion of hydrogen into helium is quite sufficient to supply all the energy requirements of a star. Energy derived from contraction is no longer needed and the star therefore ceases to contract. (We can look at it another way. The star has been collapsing inwards because there has been no way of counterbalancing the gravitational forces present. Now energy is being created at the centre of the star. This streams outwards, pushing against the stellar material and holding it up. A balanced situation is possible, with the star carefully poised in equilibrium.) This stable position can continue quite happily until all the central reserves of hydrogen have been burnt, which will take a considerable time.

We are now in a position to define the main sequence. It is that part of the H-R diagram which contains stars in their initial hydrogen-burning stage. Because stars have so much hydrogen, its burning takes up a good deal of their lifetime. As a result, a majority of all stars will be lying in this region — as is, indeed, observed.

Mass and Chemical Composition

We have noted previously that the exact track of a proto-star across the H-R diagram depends principally on its mass. The higher its mass, the higher its final position on the main sequence. Thus our theory of stellar evolution leads us to suppose that there will be a relationship between the brightness and the mass of a star on the main sequence. This is simply the mass-luminosity relation which we noted could be derived from the observations in Chapter 1. Theory takes us a little further, however. Because the exact equilibrium attained depends on the amount of hydrogen

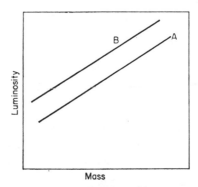

FIG. 32. Mass-luminosity relations for Population I and Population II stars. A – Population I star with 75 per cent hydrogen, 24 per cent helium and 1 per cent heavier elements. B – Population II star with 75 per cent hydrogen, 24·9 per cent helium and 0·1 per cent heavier elements.

present to be burnt, the brightness of a star of given mass depends slightly on its chemical composition. The mass-luminosity relation should hold strictly only for a group of stars with the same chemical composition. This means, in particular, that Population I and Population II stars should yield appreciably different mass-luminosity relations. As it happens, the stars with well-determined masses are those which are close to us in the sky, and they are all Population I. The mass–luminosity relation derived from observation is therefore one appropriate for Population I.

As the more massive stars are brighter and burn up their hydrogen more quickly, they remain on the main sequence for a relatively brief time. A star fifteen times more massive than the Sun will leave the main-sequence region after some ten million years. The Sun will remain on the main sequence for some ten thousand million years. (In both cases these times are considerably longer than the time required for the respective proto-stars to contract to the main sequence.) The length of time required by the Sun to use up all its central hydrogen supply is, in fact, comparable with the present age of our Galaxy. This means that stars less massive than the Sun, that were born in our Galaxy near its beginning, are still for the most part clustered near the main-sequence region of the H-R diagram. On the other hand, only a small proportion of the stars weighing $15M_\odot$ (namely those born during the past ten million years) still remain clustered near the main sequence. Earlier members of this group have long since passed into other regions of the H-R diagram. As a result, there are far fewer stars at the bright end of the main sequence than there are at the faint end. There may be hundreds of really bright stars, but there are millions of stars like the Sun of intermediate brightness. The intrinsically faint stars at the bottom end of the main sequence are very common indeed. The difference in the number of bright and faint stars is, in fact, so marked that it cannot be ascribed only to a difference in the rate of evolution. It must be that the fragmentation process by which stars are born produces preferentially several small proto-stars rather than a single big one.

The Upper and Lower Main Sequence

Although the main sequence forms one continuous line in the H-R diagram, it is convenient to divide it into two parts and to consider each separately. As we saw in Chapter 3 hydrogen can burn in either of two ways: by the proton-proton chain or by the carbon cycle. Depending on which predominates, the structure of the star is radically altered. Which produces the major share of the energy is governed mainly by the central temperature of the star,

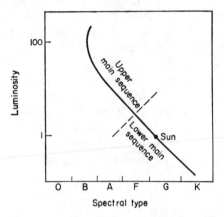

FIG. 33. Hertzsprung-Russell diagram showing the upper and lower main sequence.

and therefore by its mass. The more massive stars are hot enough for the carbon cycle to operate at their centres; the smaller stars must rely on the proton-proton chain. There is a small region of masses on the main sequence where both processes produce an appreciable proportion of the total energy; elsewhere one or the other is completely dominant. This intermediate region corresponds to stars which are about one and a half to two and a half times the mass of the Sun (for Population I). It is convenient to take this mass range as a dividing line, and to call the more massive (and, therefore, brighter) stars the upper main sequence and the less massive stars the lower main sequence.

It follows from this definition that our own Sun is a lower main-sequence star.

The Upper Main Sequence

We will consider the upper main sequence first. The carbon-nitrogen-oxygen cycle depends very sensitively on the temperature: much more so than the proton-proton chain. A star which has a central temperature twice that of the Sun produces ten thousand times more energy from the carbon cycle. But its energy production from the proton-proton chain is only greater by a factor of five. (This is why we can ignore the energy produced by the proton-proton reaction in upper main-sequence stars.) This very high temperature sensitivity has an important effect on stellar structure. When the carbon cycle is operating, the energy production increases very rapidly towards the centre of the star. There is too much energy generated for it to be efficiently transmitted by radiation. The central regions of the star therefore become unstable and most of the energy is transmitted by convection currents. On the other hand the calculations show that the outer zone of such star is stable — the energy is transferred purely by radiation. It is customary to call the inner zone of a star its *core*, and the outer zone its *envelope*. The structure of an upper main sequence star can then be summed up as consisting of a convective core and a radiative envelope.

This structure plays a fundamental role in the future evolution of the star. The rate of hydrogen-burning increases very rapidly towards the centre, so that ordinarily the central hydrogen content would be consumed in only a short space of time. But the presence of convection currents means that the central material is constantly being changed. New material is brought in from the outer parts of the core where burning is slow, and the material at the centre, which was burning rapidly, is taken out to cooler regions. This continual stirring produces a well-mixed core in which the hydrogen content decreases at the same rate throughout. In particular, the hydrogen disappears altogether from the whole core at one unique instant. The envelope of the star, even on its inner edge, is hardly hot enough for appreciable hydrogen burning; so the outer regions retain virtually their original chemical composition.

What difference does hydrogen-burning make to an upper main-sequence star? Surprisingly little at first. In the H-R diagram, the star remains quite close to its original position on the main sequence until a very considerable amount of the hydrogen in its core has been burnt. Then, gradually, it becomes slightly brighter and slightly cooler (which is equivalent to saying that it begins to increase slowly in size). It moves away from the main sequence — upwards, and to the right in the H-R diagram — but not to a very great extent. Even when all but 10 per cent of the hydrogen in the core has been consumed, such a star may not be much more than twice as bright as it was initially, and its surface temperature may not have dropped by more than a tenth.

When the upper main-sequence star has completely exhausted its central hydrogen content, it is left with no sufficient nuclear energy source immediately available. It therefore undergoes internal changes which actually take it well away from the main sequence in the H-R diagram. We will take up consideration of these changes in Chapter 7.

For the moment, we will concentrate on the hydrogen-burning period. It is important to note that a star stays near the main sequence during this period, but not exactly at the spot where it first arrived. Since the main sequence represents stars covering a considerable range of ages, it does not appear as a narrow line in the H-R diagram, but as a band of appreciable width. Astronomers usually distinguish between the zero-age main sequence and the observable main sequence. The former is the line in the H-R diagram which corresponds to the positions of the stars at the instant they start their main-sequence career. The latter corresponds to the main sequence derived from surrounding field stars (which are of widely differing ages). We can obtain a near approximation to the zero-age main-sequence, in practice, by examining the H-R diagrams of open clusters, where all the stars are of similar age. No single open cluster will show a complete zero-age main sequence for all masses of stars. In a very young cluster (say a million years old), the brighter, more massive, stars will have had time to contract to the main sequence, but the fainter stars, which take longer, will still be contracting. They will therefore lie to the right of, and above, the zero-age main sequence. On the other hand, in an older galactic cluster (say a hundred million years old) the fainter stars will have had time to contract to the main sequence, but the brighter stars will have consumed their central hydrogen supply and moved away. To check our theoretical calculations of the position of the main sequence, we must therefore combine the observed H-R diagrams of both old and young clusters.

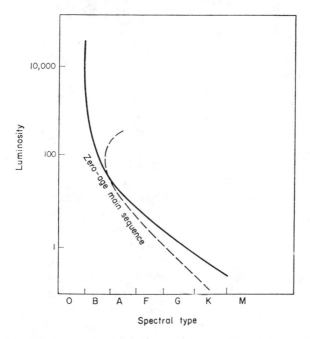

Fig. 34. The zero-age main sequence as deduced from old (– – – – – –) and young (————) clusters.

Our description so far has slightly over-simplified what happens within an upper main-sequence star. Detailed calculations show, in fact, that the core and the envelope of such a star do not remain completely separate. As the convective core uses up more and more of the central hydrogen content it also decreases in size. During the retreat inwards it leaves behind a zone which was previously transferring energy by convection but is now stable: passing energy by radiation only. This newly created radiative zone was previously subjected to hydrogen-burning; but, during

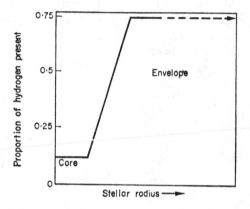

Fig. 35. Decrease in the hydrogen content from the surface to the centre of an upper main-sequence star.

the retreat of the core, the temperature drops below the ignition point. As a result, this intermediate zone has less hydrogen content than the original radiative envelope, but more than the still burning core. The structure we must now envisage contains therefore three regions, rather than two: an outer radiative envelope, an intermediate radiative zone and an inner convective core. The part of the intermediate zone near the envelope will have a hydrogen content which is only slightly diminished because it will only have been in the hydrogen-burning core for a brief period of time. But the part of the zone near the core will have much less hydrogen than the outer regions because it has been burnt for much longer. The overall effect is to build up a gradient in the chemical composition of the star between the envelope and the core (Fig. 35).

This difference in structure actually has very little effect on the track of the star in the H-R diagram. Our previous discussion still holds in all essentials. There is a slight change right at the end of the hydrogen-burning period. The central temperature rises very rapidly as the hydrogen in the core approaches exhaustion. The surface temperature of the star therefore also rises at this point and the star moves back a little towards the left in the H-R diagram. When the final remnants of the hydrogen in the core have been consumed, the surface temperature decreases once more, and the star moves back towards the right. In the H-R diagram this part of the star's track is an S-shaped bend which accelerates abruptly as the star finally consumes all its central hydrogen and begins to evolve rapidly.

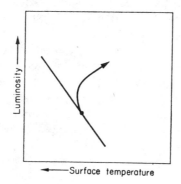

FIG. 36. The last stages of hydrogen burning in an upper main-sequence star. (A small segment of the upper main sequence is also shown.)

The Upper Limit of the Main Sequence

The main sequence does not continue upwards indefinitely: there is an upper limit to how bright and hot a main-sequence star can be. We must obviously enquire what sets this limit. We have seen that, in a stable star, the gravitational pull inwards on the stellar material is balanced by the opposing push outwards of the pressure. The pressure is made up of two components — one from the high temperature of the atomic particles present, and the other from the radiation streaming outwards from the centre. In stars like the Sun, the first component — the gas pressure — is much more important than the second — the radiation pressure. In a very massive star, however, the radiation pressure becomes large because the amount of radiation produced in the central core is large. This pressure opposes the gravitational pull inwards of the star. Indeed, at a certain mass, the outward radiation pressure becomes equal to the inward gravitational attraction. This obviously forms a limit beyond which a star cannot hold together. Corresponding to this limiting mass, there is an upper limit to the brightness and surface temperature of a main-sequence star. In fact, the limitation on the mass is likely to occur before the main-sequence stage, whilst the star is still forming. The amount of extra material that can be accreted to a proto-star core will depend on the radiation pressure from the part of the core that has already formed. Theoretical calculations place the limit at not more than $100M_\odot$. The most massive stars so far observed are the two components of a binary (known, after its discoverer, as Plaskett's star). They both probably weigh some $70M_\odot$. It is interesting that these stars give indications of being unstable and seem to be losing material into space.

The Lower Main Sequence

We can now examine the development of stars on the lower main sequence (that is stars of $2M_\odot$ or less). The energy of these stars is due almost entirely to the proton-proton chain. The reactions in this chain are much less sensitive to the temperature than those in the carbon cycle. As a result, the energy production builds up less rapidly towards the centre of the star and the stellar material does not become convectively unstable. Hence, energy transport in the central regions of a lower main-sequence star is by radiation and these regions consequently remain unmixed. In the outer parts of one of these stars the conditions again differ from the upper main sequence. The envelopes are at appreciably lower temperatures, which makes them considerably more opaque to the passage of radiation than the envelopes of upper main-sequence stars. Indeed, they dam back the radiation flux to such a degree that their stability breaks down and the main energy flow is carried by convection currents. Thus the internal structure is completely different for stars on the upper and lower main sequence. The former transmit energy by convection in the core and radiation in the envelope; the latter by radiation in the core and convection in the envelope.

The determination of the exact conditions in a convective envelope is one of the most difficult problems in the current theory of stellar evolution. To see why, we must look in more detail at the process of convective energy transfer. We can imagine that it operates in the following way. At some level in the star a bubble collects which has more energy (in the form of heat) than its surroundings. Because it is hotter, and therefore lighter, it rises towards the surface of the star. After moving upwards for a certain distance (which will depend on the exact conditions in this region of the star) the bubble stops, giving up its heat to its new surroundings. The distance the bubble moves between picking up its excess energy and releasing it again is called the *mixing length* of that bubble. Different bubbles will rise different distances before disappearing, but it should be possible for a given set of circumstances, to define an average mixing length for the material.

Unfortunately, there is no adequate theoretical method available for estimating this mixing length. Nor, since the convection is occurring in the stellar interiors, can we hope to measure this quantity. A rough calculation shows that the mixing length varies tremendously with the conditions. The mixing length in the core of a star is extremely large. As a result, the exact value used in the calculations for this region is unimportant: it makes no significant difference to the answers obtained. But in the envelope of a star, particularly near the surface, the mixing length is small. Even quite minor changes in the choice of length can lead to considerable variations in the calculated structure. In terms of stellar models, we can say that the conditions in a convective core can oe calculated accurately, but the conditions in a convective envelope cannot. This means, in turn, that the structure of lower main-sequence stars cannot, at the moment, be worked out with the same accuracy as the structure of upper main-sequence stars.

Since the central regions of a lower main-sequence star are stable, the helium produced by hydrogen burning stays where it is formed. As the temperature of the star increases towards the centre, so does the rate of hydrogen burning, and the hydrogen

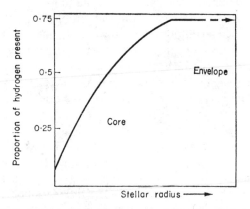

FIG. 37. Decrease in the hydrogen content from the surface to the centre of a lower main-sequence star.

supply at the centre disappears more rapidly than that further out. The evolving star therefore develops a variable chemical composition. From the surface to the edge of the core it remains the same, but further inwards the amount of hydrogen present at any given time decreases.

As the internal composition changes, so do the characteristics of the star. It begins to creep gradually up the main sequence.

This means that its size remains much the same, but its surface temperature and brightness both gradually increase. This differs from the evolution of upper main-sequence stars, which increase slightly in radius as hydrogen burning proceeds. Both types of star are similar, however, in that they remain close to the main sequence. We have seen that rapid evolution of an upper main-sequence star starts when all the central hydrogen is burnt to helium. This occurs simultaneously throughout the core. In a lower main-sequence star the complete conversion of hydrogen to helium in the core is not instantaneous. The hydrogen is used up first right at the centre of the star and then subsequently at points further out. However, we can consider the instant when the hydrogen at the centre is completely burnt as being rather similar

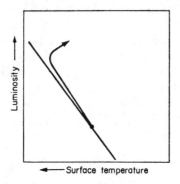

Fig. 38. The last stages of hydrogen burning in a lower main-sequence star. (A small segment of the lower main sequence is also shown.)

to the conversion of the entire core in an upper main-sequence star. There are two reasons for this. The first is that the complete conversion of the entire core to helium in a lower main-sequence star follows fairly soon after the initial exhaustion at the centre. The second reason is that at this point the lower main-sequence star stops moving up the main sequence and commences moving to the right instead (which means that its radius starts to increase).

It is now apparent that, although the internal structures of upper and lower main-sequence stars are quite different, from the observational standpoint their initial development is remarkably similar. Both stay in the general region of the main sequence until they have exhausted their central hydrogen supplies, then they move off towards the right. The big difference observationally is their rate of evolution. This is partly because lower main-sequence stars, being of less mass, necessarily evolve more slowly than upper main-sequence stars, and therefore remain much longer in the general vicinity of the main sequence. But there is also the factor that the upper main-sequence stars exhaust their central hydrogen supplies instantaneously and so move on suddenly to their next evolutionary phase, whereas the lower main-sequence stars, burning more gradually, move on more slowly. This difference affects the later stellar evolution which we will take up in Chapter 7.

Since the initial evolution of lower main-sequence stars bears a superficial resemblance to that of upper main-sequence stars the comments previously made about the finite width of the main sequence still hold. As we have seen before, the position of the zero-age position for the lower main-sequence can be determined by studying the H-R diagrams of fairly old open clusters.

The Lower Limit of the Main Sequence

Just as the main sequence has an upper limit in the H-R diagram, so it has a lower limit. There is a certain brightness and surface temperature below which no main sequence stars have been found. We could predict that such a limit would occur from our idea that stars form by contraction. We have seen that the rate of contraction falls off rapidly for the less massive stars. There must be some mass of star which contracts so slowly that it will not yet have had time to reach the main sequence even though it started contracting when the Galaxy itself first formed. We can, however, adduce an even more fundamental reason why the lower main sequence should have a definite end. We have

defined the main sequence as that part of the H-R diagram where
stars burn hydrogen in their cores. If hydrogen is to start burning,
a certain minimum temperature (over a million degrees) must be
attained at the centre of the star. The temperature actually
reached during the contraction process depends on the mass of
the star. As we go to stars of lower and lower mass, so the
maximum attainable central temperature decreases. Eventually
we reach a mass which is so small that the central temperature
never rises high enough to ignite the hydrogen present. If a
proto-star develops with this mass, or less, it can never become
a normal main-sequence star; so this mass represents a lower
limit to the extension of the main sequence.

Chemical Composition
and the Evolution of Main-sequence Stars

So far we have been considering the mass of a star as the only
thing influencing its evolution. This is a reasonable assumption
when dealing with a homogeneous group of stars such as a star
cluster. But in the Galaxy as a whole many different chemical
compositions occur — ranging from extreme Population I to
extreme Population II. The discussion of main-sequence stars
in this chapter has been mainly concerned with Population I
stars, but most of it is equally applicable to Population II
stars.

We have seen that Population II stars have appreciably less of
the heavier elements than Population I. On the other hand, they
may possibly have more hydrogen. Thus a Population I star
might be 70 per cent hydrogen, 27 per cent helium and 3 per
cent heavier elements. A Population II star of the same mass
might contain less than 0·1 per cent of heavier elements, but it is
just possible that the hydrogen content might be as high as
90 per cent. These differences in chemical composition are
reflected in the different brightnesses and surface temperatures of
the stars. For a star of given mass, the less hydrogen present, the
hotter and brighter it is. Similarly, a decrease in the amount of

heavy elements present makes a star hotter and brighter. Thus the brightest main-sequence star of a given mass is, paradoxically enough, the one which contains the most helium.

On a mass-luminosity diagram these differences between Population I and II stars show up clearly. They are by no means so obvious in an H-R diagram. Since the brightness and surface temperature both increase in going from Population I to Population II stars, changes in chemical composition move a star along the main sequence rather than off it to the left or right. For this reason, the main sequence corresponding to field stars (which may include stars of widely varying chemical composition) does not have a unique mass of star for each value of the brightness and surface temperature. A star at any given point on such a main sequence can have a range of masses depending on its precise chemical composition. For example, a star of solar mass with a hydrogen content of 60 per cent may have the same brightness as a star of $1 \cdot 3 M_\odot$ and a hydrogen content of 80 per cent. The difference in composition also means that the dividing line between the upper and lower main sequence (drawn in terms of the predominant energy-generation mechanism) is not the same for Population I and II stars. For the former, as we have seen, it lies in the region $1 \cdot 5 - 2 \cdot 5 M_\odot$. For the latter, it is at about $4 M_\odot$.

Main-sequence Stars of Intermediate Mass

The study of stellar models with masses in the intermediate range (say $1 \cdot 5 - 2 \cdot 5 M_\odot$ for Population I), where both the proton-proton chain and the carbon cycle are active, presents certain additional difficulties. Such a star will have a small convective core at the centre burning hydrogen by the carbon cycle. Round this there will be a stable region where hydrogen is burning by the proton-proton chain. These two regions together correspond to the energy-producing core of the star in our previous formulation. Next there will be a radiative region, where no energy is generated, and, finally, a convective zone extending up to the stellar surface. These latter two regions may be called the stellar envelope. In other words, the structure of stars in this mass range represents a combination of upper main-sequence and lower main-sequence characteristics. Similarly, the evolution of these stars is a compromise. They move partly up the main sequence and partly to the right.

The Age of Star Clusters

We have noted that one of the main obstacles to studying stellar evolution is our lack of knowledge of stellar ages. We may suppose that all the stars in a star cluster have the same age, but we cannot say immediately what that age is. We can guess, however, that the stars will also all have similar chemical compositions. (This can, of course, be checked by observation.) Then the evolutionary differences between the cluster stars will depend only on their masses. This is a particularly simple case to study.

We calculate models of main-sequence stars which have the same age and the same initial chemical composition, but different masses. We plot the positions of our calculated model stars in an H-R diagram. We will refer to this as a *theoretical* H-R diagram. We then resume our calculations and plot an equivalent diagram for the same group of stars when they are slightly older. We repeat this process until we have a collection of theoretical H-R diagrams each of which represents the stars at a specific age. Finally, we compare all our theoretical H-R diagrams with that obtained from observing an actual star cluster (so long as it has the same chemical composition). We choose the theoretical diagram that corresponds most nearly to the observational diagram and read off the age. This represents the theoretical age of the cluster.

Since star clusters occur throughout the whole of our Galaxy, we can hope to discover in this way which are the older parts of the Galaxy, which the younger, and what sort of ages are involved. This is vital information for investigating the origin and evolution of our Galaxy as a whole.

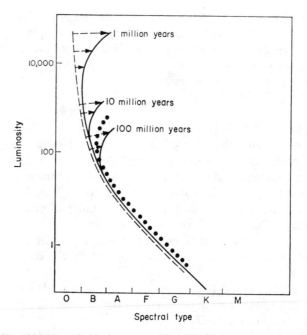

FIG. 39. The age of a star cluster derived from stellar models.
– – – – – – zero-age main sequence. ———— theoretical tracks
for star clusters with the corresponding ages. · · · · · · · track of
an observed star cluster: we can deduce from the diagram that it is
more than 10 million years old and less than 100 million years.

Mixing in Stars

We have assumed, when discussing the evolution of main-
sequence stars, that mixing only occurs when energy is transmitted
by convection. We have supposed that radiative transfer implies
an absolutely immobile stellar material. What if this assumption
is untrue: would it make much difference to the evolution of a
main-sequence star?

If zones in radiative equilibrium mixed material just like
convective zones, then the whole of a main-sequence star, whatever its structure, would be mixed together. The main-sequence

stars we have previously described evolved with a core deficient in hydrogen but their envelope retained the primeval amount. A thoroughly mixed star would use up hydrogen uniformly throughout its volume. (Just as the convective core of an upper main-sequence star uses up hydrogen uniformly.) This suggests immediately that a mixed star would take longer to evolve: because it has more hydrogen available at the centre for burning. If we study how a main-sequence star like the Sun would evolve under conditions of complete mixing, we find that it becomes initially hotter and brighter, and moves up the main sequence in much the same way as when the core and the envelope are taken to be distinct.

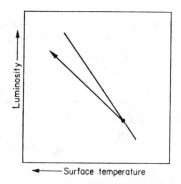

FIG. 40. The initial evolutionary track of a completely mixed main-sequence star. (A small segment of the main sequence is also shown.)

The difference comes when most of the hydrogen present has been consumed. Then a completely mixed star, instead of evolving off to the right of the main sequence, moves off to the left. Now we can be sure from our study of the H-R diagrams of star clusters that most stars do, in fact, leave the main sequence by moving to the right. This means that our first assumption — that mixing only occurs in convective zones — is supported by the observational evidence. We cannot exclude, however, the possibility that a few main-sequence stars may be mixed sufficiently to move off to the left. It is worth noting, too, that the very low-mass stars at the bottom of the main sequence are completely mixed. For these stars the outer convective envelope has extended so far that it reaches right down to the centre. We may expect that the evolutionary development of these stars will differ from that of stars further up the main sequence.

Rotational Mixing

Is there any mechanism, other than convection, which might be capable of mixing the contents of a star? Spectroscopic studies of stars have shown that they spin round on their axes just as the Earth does. It has been found that the spin rate depends on the type of star observed. Stars on the upper main sequence (the O, B and A stars in the Harvard classification) are generally spinning quite fast. The spin rate falls off through the F stars and is too low to be measured for later-type stars. This means that all lower main-sequence stars (such as the Sun) spin rather slowly. Thus a point on the Sun's equator moves round with a speed of about one mile per second (not much larger than the speed at the Earth's equator). But a point on the equator of a bright star may be moving at 250–300 miles per second.

A slow spin rate has no effect on the evolution of a star. It has been calculated, however, that high rates of spin can produce currents in the stellar material which will move from the centre of the star to the surface and back again. Such currents might be capable of mixing together material from the core and the envelope. Fortunately, more detailed calculations show that the time required for rotational mixing to become effective is usually long compared with the lifetime of a star on the main sequence. Nevertheless, the calculations are difficult to make, and there are complicating factors present — such as magnetic fields — whose effect is still uncertain. It is therefore still possible that very fast rotating stars (which means, in practice, a few of the upper main-sequence stars) may evolve to the left of the main sequence instead of to the right. The total number that do this must certainly be small. No other method of mixing is likely to be more efficient than rotation (although diffusion — the gradual inter-mixing of different gases when they are in contact with each other — may be important under some circumstances) so we can safely conclude that the vast majority of main-sequence stars evolve to the right as has been previously described.

Observational Indications of Mixing

Can we find any stars in the sky which provide observational evidence of mixing? The most interesting possible suspects are the Wolf-Rayet stars (named after the two astronomers who first discovered them). They are rare stars, bright and very hot, which tend to be found in association with young O and B stars. They thus lie high in the top left-hand corner of the H-R diagram — round about the top of the main sequence. But they are clearly not normal main-sequence stars. For one thing, matter is streaming out from their surfaces into space at very high speeds. For another, they have very peculiar spectra. The Wolf-Rayet stars found in associations have an extraordinarily large amount of nitrogen present in their atmospheres. This can be interpreted as signifying that they are massive stars which have passed rapidly through the stage of hydrogen burning by the carbon cycle. Subsequently, the burnt material from their cores has been mixed in with the unburnt envelope material. The argument runs as follows. Although the carbon, nitrogen and oxygen nuclei are used as catalysts in hydrogen burning, and therefore are not destroyed in overall amount, nevertheless they are transmuted into each other. Due to the different rates at which the various reactions in the carbon cycle go, carbon and oxygen are trans-muted preferentially into nitrogen. Suppose, for example, that when a star starts burning hydrogen by the carbon cycle it has the following proportion of nuclei per unit volume of the core: 100 carbon nuclei, 100 oxygen nuclei and 100 nitrogen nuclei. After the carbon cycle has been operating for a while these proportions will have changed to: 3–4 carbon nuclei, 1–2 oxygen nuclei and

295 nitrogen nuclei. Thus the total number of the three nuclei present remains the same, but their relative proportions are considerably altered during hydrogen burning. If, then, the central core of such a star is mixed with its outer atmosphere we will see a large increase in the proportion of nitrogen present.

Supposing we accept that Wolf-Rayet stars are mixed, can we point to any mechanism which might cause the mixing? Many Wolf-Rayet stars have been found to be members of close double stars. It is simplest therefore to suppose that mixing is brought about by some form of gravitational interaction with the other star.

Mass Change and Evolution

We have made one further implicit assumption in deriving our picture of the initial evolution of main-sequence stars. We have supposed that the star always has the same mass. We must now enquire how a gain or a loss of material would alter the development.

It seems unlikely that accretion of material from space is of major importance for stellar evolution after the initial process of star formation. (The important exception is when two stars are close together in a binary system.) On the other hand, we can find plenty of observational evidence to indicate that mass loss occurs from stars at various stages of their lives. This is true of stars at the beginning of their careers (T Tauri) or towards their end (red giants). However, we have no need to look so far afield for indications of mass loss. The Earth itself is bathed in the material of the solar wind which streams outwards from the Sun. (Although the solar wind has important effects for us on Earth, the actual amount of material being lost by the Sun is infinitesimal compared with the total solar mass. It has no significant effect on the future evolution of the Sun.)

When we examine other main-sequence stars, we find that mass loss seems to be greatest from the bright, hot stars at the top of the main sequence. We have seen that this is true of Wolf-Rayet stars, but these may not be normal main-sequence stars. Another obvious example is the Be stars where we can actually observe the

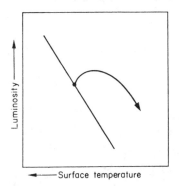

Fig. 41. Initial evolution from the upper main sequence for a star losing mass. This should be compared with Fig. 36. (A small segment of the upper main sequence is also shown.)

material being ejected. Unfortunately, it is difficult to estimate the rate of mass loss in Be stars as they eject material irregularly, often with long periods between the successive bursts. The rough estimates available suggest that the rate may just be significant for the evolution of the star.

Since the brightness of a main-sequence star depends on its mass, removal of mass diminishes the brightness. If a bright star on the main sequence loses mass at an appreciable rate its initial evolution from the sequence will be altered. Instead of following a horizontal or a slightly upward track to the right in the H-R diagram, it will move downwards — the exact angle depending on the rate of mass loss.

There are certain main-sequence stars for which mass loss almost certainly plays an important role. These are the components of very close binaries — sometimes called contact binaries. The stars are so close together that they have very strong gravitational interactions. Material can be removed from their surfaces, to float in the surrounding space, or even to dissipate completely. This very greatly affects their post main-sequence evolution as will be described in Chapter 8.

CHAPTER 6

The Sun as a Star

THE Sun is so important to us that we tend not to think of it as just a typical star. Yet its characteristics are perfectly normal: in size, mass and brightness it lies fairly near the middle of main-sequence stars. (Since small, faint stars of low mass are much commoner than bright massive stars, the Sun's characteristics actually place it rather above the average, if we are concerned with the relative numbers of the different types of star.) The major significance of the Sun for our general studies of stars lies in its closeness to us. It is the only star whose surface can be examined in detail. We can, for example, analyse the structure of the solar atmosphere in a way that is quite impossible for other stars. However, having established that the Sun is a typical star, we can use our detailed knowledge of its structure to elucidate what is happening in other stars.

The Evolution of the Sun

The Sun, as a normal, lower main-sequence star, has presumably evolved in the manner described in the last chapter. It long ago reached the main-sequence by contraction, and is now almost immobile in the H-R diagram, consuming its central hydrogen reserves. Our knowledge of the Sun's evolution is, however, in one respect much more complete than for other stars: we can specify a lower limit for the length of time the Sun has been on the main sequence. From a study of terrestrial rocks and, even more, from a study of meteorites (some of which may be primeval material left over from the formation of the solar system) we can date the origin of the planets to between 4–5 thousand million years ago. All our current ideas suggest that the Sun was formed at the same time as the planetary system; so this is also the probable age of the Sun.

Far more stellar models have been calculated for the Sun than for any other star. Because our knowledge of the Sun — its size, mass, etc. — is so precise, our models must be very accurately designed to give a good fit. The Sun has, indeed, come to be regarded as a standard against which theories of stellar evolution may be measured: if we can explain the present characteristics of the Sun accurately, then maybe our theories will work fairly well for other stars about which we know less. When the past evolution of the Sun is worked out along the lines discussed in the previous chapter we find that it has moved appreciably from its zero-age position on the main sequence. It has already consumed a considerable quantity of the hydrogen in its core — perhaps only a third of the original amount is now left at its centre. As a result, the Sun may have brightened by as much as 50 per cent since the time it was first formed. This brightening will continue gradually until all the central hydrogen has been burnt. Then there will begin a rapid expansion. Well before this, however, life on Earth will have become unsupportable. As the Sun brightens, the surface temperature of the Earth increases. By the time the Sun has finished off its central hydrogen, the Earth's temperature will be above the boiling point of water. However, we need fear no immediate catastrophe; the Sun is still only half-way through its life on the main sequence. Five thousand million years may have elapsed since it first burnt hydrogen, but there is probably a similar length of time still to go before the central reserves are completely depleted.

Deuterium, Lithium, Beryllium and Boron

The very precise models which have been calculated for the Sun also throw light on its earlier evolution, when it was contracting to the main sequence. As a solar-type star passes through its final stages of contraction, the convection currents which formerly traversed the whole star are gradually confined to the outer layers (which we have previously called the envelope). Convection currents, we know, imply mixing. Whereas during the latter part of the contraction the whole of the star intermixed, in the final main-sequence Sun this was only true of the outer layers.

We have seen that certain light nuclei — deuterium, lithium, beryllium and boron — burn very easily. If we compare the amount of these elements on the surface of the Sun with the amount present on Earth, we find that beryllium and boron seem to occur to about the same extent in both, but that the Sun has much less deuterium and lithium than the Earth. The most probable explanation of this difference depends on the exact structure of the Sun during its final contraction stage. Laboratory experiments show that deuterium and lithium transmute at somewhat lower temperatures than beryllium or boron. We must assume that while the contracting Sun was still completely convective its central temperature reached a value high enough to ignite deuterium and lithium but not beryllium and boron. At this stage in the Sun's existence all of its substance is passing through the central regions and so being processed. Deuterium and lithium will therefore be rapidly depleted throughout. The proto-Sun contracts further until the central temperature has risen high enough to burn beryllium and boron. But now there is no longer complete convection: the central zone has become stationary. Mixing of core with envelope has therefore ceased and the beryllium and boron in the latter region never reach the hotter central parts. Our current observations of the Sun will then, indeed, show more beryllium and boron present than deuterium and lithium.

We can use this observation to check our theoretical ideas on the depth of the solar convective envelope. It can extend down to depths with a temperature high enough to burn lithium slowly, but not so far that beryllium begins to burn. The depth determined from this criterion agrees quite well with that derived from theoretical models. The same observational test can be applied to other stars.

We find that stars like the Sun all seem to be deficient in lithium but not in beryllium. Obviously all solar-type stars have similar convective histories. On the other hand, the T Tauri stars, which, as we have seen, are probably still in the contraction stage, have an amount of lithium in their atmospheres similar to the amount on Earth. They must be stars which have not yet heated up sufficiently to burn lithium. Presumably if we could watch them over a long period of time, we would observe the proportion of lithium present to decrease as they approach the main sequence. The theoretical models suggest that upper main-sequence stars, due to their different structure, should also retain the primeval amount of lithium in their surface layers. This, too, seems to be confirmed by observation.

The Rotation of the Sun

We have noted that the Sun rotates very slowly. Indeed, if we were not very close to it, the rotation rate would be too low to be measured. It was suggested in Chapter 4 that most of the spin has been siphoned off by magnetic fields during the formation of the Sun. We can try to test the validity of this explanation by observing the current state of the solar system. We can, for example, observe the present solar magnetic fields, see how far into space they stretch and investigate their interaction with the interplanetary material.

It is tempting to suppose that all lower main-sequence stars like the Sun have evolved in the same way with a loss of rotation to the surrounding medium. In other words, stars like the Sun should all have the chance of developing planetary systems. On the other hand, rapidly spinning, upper main-sequence stars must be supposed to transmit little of their energy to the surrounding medium (perhaps because they have heated the material so much that it expanded rapidly and blew away). Brighter stars would therefore have had little opportunity of acquiring planetary systems. This may well be the true interpretation of the picture, but it is certainly interesting that the division between rapidly spinning and slowly spinning stars occurs at very nearly the dividing point between the upper and the lower main sequence. Rapidly rotating stars are also those stars with a convective core and radiative envelope; slowly rotating stars are also generally those with a radiative core and a convective envelope. We cannot yet entirely rule out the possibility that the difference in spin rate is due to the difference in structure rather than the presence, or absence, of a planetary system.

The Solar Magnetic Fields

A magnetic field streams out from the Sun just as it does from the Earth, or from an ordinary bar magnet. However, the solar magnetic field bends over towards the equator a good deal more sharply than would be expected (Fig. 42). This field is called the Sun's *poloidal* field, because it seems to emanate from two

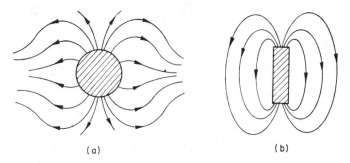

(a) (b)

FIG. 42. The poloidal magnetic field of the Sun (a) compared with the field of a bar magnet (b).

magnetic poles within the Sun. The name also serves to distinguish it from the other much more intense magnetic field — called the *toroidal* field — which is found round the Sun's equator. A very high proportion of all the changes observed in the solar atmosphere are due to the activity of this toroidal field. For example, the dark patches known as sunspots occur where loops of the toroidal field intersect the solar surface. Sometimes this intense field becomes highly distorted. Then there arises the possibility that the magnetic field will squeeze and compress adjacent regions of the solar atmosphere. This may produce considerable local heating which leads, in turn, to the emission of much X-ray and ultra-violet radiation and the ejection of high-energy atomic particles. It is generally supposed that the poloidal magnetic field is the

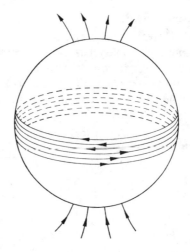

Fig. 43. The toroidal magnetic field of the Sun.

more fundamental and that the toroidal field is somehow derived from it. Various mechanisms have been suggested, but for our purpose it is sufficient to know that the solar magnetic fields can be reasonably explained in terms of star formation by contraction. (It is a simple matter to show that the weak galactic magnetic field found in interstellar clouds may be compressed by just the right amount during proto-star contraction to produce the present solar poloidal field.)

Stellar Magnetic Fields

Our study of the Sun suggests two questions to be asked about other stars. Do they have magnetic fields like the Sun? If so, do these magnetic fields produce violent activity such as we see on the Sun? The presence of magnetic fields is well established, in a wide variety of stars. The strongest magnetism has been found in white dwarfs (up to a thousand million times as strong as the Earth's magnetic field). So far as main-sequence stars are concerned, most attention has concentrated on the Ap stars (with magnetic fields which may reach nearly a hundred thousand times the intensity of the Earth's field). The fact that this group of stars, in particular, is distinguished by strong magnetic fields ought to give us some indication of how magnetic fields in stars are generated. The problem is to decide which of the characteristics of Ap stars are the cause of magnetic fields, which are, instead, caused by the magnetic fields, and which are completely unconnected with magnetic phenomena at all. For example, the peculiar chemical composition of Ap stars — which was the initial reason for their segregation as a separate group — is generally agreed to be a result of the magnetic fields present; though the exact process which operates is still disputed. It has been suggested that there are two essential properties for the production of magnetic fields which must be possessed by Ap stars. The first is that they must be spinning rapidly on their axes, the second that they must have thin convective zones just below their surfaces. (This latter requirement differentiates them from hotter stars, which have no convective zone at all, and fainter stars which have a much thicker convective zone.) According to this speculation, when these two properties come together in the same star, large magnetic fields must inevitably be generated. One feature of the magnetic fields in Ap stars is that they frequently undergo rapid variation. Observations of several Ap stars suggest that their magnetic poles reverse their positions every few days (the north magnetic pole turns into the south magnetic pole and *vice versa*). It is interesting that the magnetic field of the Sun reverses in the same way, but over the much longer period of eleven years. According to the geological evidence, the Earth's magnetic field, too, has reversed in times past, but only at intervals of several thousand years.

In groups of stars other than the Ap stars there does not seem to be a general tendency for large magnetic fields. A few members of most types of stars have significant magnetic fields. We cannot say that the others possess no field, but only that it is too small to be measured by our present instruments. We find that some very hot stars have magnetic fields, so do some quite cool stars; rapidly rotating stars have fields, but so do some slow rotaters. The observations suggest that the conditions required for the production of medium-intensity magnetic fields are not very stringent.

Stellar Activity

Since magnetic fields on the Sun cause solar activity, we might enquire whether, in an analogous way, stellar magnetic fields are linked with stellar activity. First of all, if we saw the Sun from a great distance, rather than from close to, would solar activity be identifiable? The answer is almost certainly no. Even when there are a maximum number of spots on the Sun's surface, its brightness is virtually unchanged. Nor can the largest flares produce a significant increase in the solar luminosity. The charged, accelerated particles produced by solar flares would also be undetectable at large distances. We might hope that stars with greater magnetism than the Sun would produce more obvious effects. This obviously means looking first at the Ap stars. But here, too, we are disappointed. The great and rapid changes in magnetic field seem to have little effect on the light emitted. Nevertheless, very small variations in the light, in step with the magnetic variations, have been detected. We may tentatively consider this as a form of stellar activity.

Much more certain and impressive evidence of stellar activity is provided by the flare stars. These are stars much fainter and cooler than the Sun which blaze up briefly at intervals. There are good observational reasons for supposing that this flare phenomenon is directly related to similar activity on the Sun. For example, solar flares produce radiation at all frequencies. This means that besides emitting much light, they also produce strong radio signals. Simultaneous measurements by optical and radio telescopes have shown that, when flare stars erupt and become much brighter visually, they also emit a sudden burst of radio energy just as solar flares do. These flare stars are above the lower main sequence — presumably still in the contraction stage. They seem, indeed, to be near the region of the H-R diagram where contracting stars change over from being completely convective to having a radiative core. It has been suggested that flaring may be a method whereby such a star rids itself of excessive rotation and magnetic field.

Magnetic Fields and Evolution

We are led on to enquire whether the presence of a magnetic field makes any difference to the evolution of a star. Our previous discussion of the formation of lower main-sequence stars suggests that it may make a considerable difference for them. If we consider the stars after they have reached the main sequence, however, we do not seem to find any great structural differences between stars of the same type with large and small magnetic fields. Again, we would expect the Ap stars to show the greatest differences, but, even for these stars, evolutionary effects are not obvious. If the magnetic field pervaded the whole star, it might prevent the normal separation of the star into a distinct core and envelope. In this case, the star would evolve rather like the completely mixed stars we have described before: to the left of the main sequence instead of to the right. But if, as seems more likely, the magnetic field is mainly confined in the envelopes of Ap stars, then its effect on evolution will be greatly diminished.

One part of the evolutionary track where our ideas might be confused by the presence of large magnetic fields is in the region of the pulsating variables. As will be explained in a subsequent chapter, brightness changes in RR Lyrae stars may provide us with information about the later stages of stellar evolution. The mechanism which produces the pulsation is located in the envelopes of these stars. If there is a large magnetic field present in the envelope, the way in which the star pulsates may be changed. Our evolutionary investigations would then be thrown out of gear. It should be added that the prototype star of this group — RR Lyrae — has, itself, been found to possess a significant magnetic field.

Solar and Stellar Coronae

The solar atmosphere extends much further into space than would be predicted *a priori*. The probable reason for this has only been worked out during the past few years. We have seen that the outer envelope of the Sun transmits energy mainly by convection. In effect, a constant stream of bubbles of the material rises towards every point on the solar surface. As a result, the bottom of the solar atmosphere is set in a state of continual vibration by the continual pushes it receives from below as successive bubbles arrive at the surface. These vibrations are transmitted upwards into the outer atmosphere of the Sun. Their speed is dependent on the density of the atmosphere: the more rarified the atmosphere, the faster they move. Now the solar atmosphere, like the Earth's atmosphere, becomes more rarified with height. This means that, as the vibrations from the Sun's surface move upwards into the higher parts of the solar atmosphere, they continually move faster. Eventually their speed exceeds the speed of sound in the atmosphere at that point. Then, suddenly, the vibrations change their nature: they transform into shock waves. (We are well acquainted with shock waves nowadays. When a jet plane flies past at less than the speed of sound we hear its engine in the normal way, but when it passes at supersonic speed we hear, instead, an abrupt bang as its shock wave hits us.) There is a vital difference between subsonic and supersonic waves. A subsonic vibration loses relatively little of its energy as it moves along. A shock wave, on the contrary, gives up energy quite rapidly to its surroundings (and thereby, of course, rapidly dissipates itself). What happens in the solar atmosphere, therefore, is that a vibration started off from the solar surface moves upward until at some height it transforms

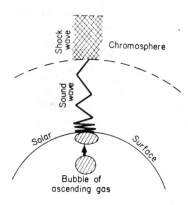

FIG. 44. Initiation of a shock wave in the solar atmosphere.

into a shock wave. From that point onwards it begins to pump energy into its surroundings. This energy heats the upper atmosphere of the Sun and causes it to expand. Thus we can explain its excessive extension. The region where the vibrations change into shock waves is the solar chromosphere. The expanded region of the atmosphere above it is the solar corona.

The solar corona extends far out into space beyond the Earth. The heating which causes its expansion is such that the coronal material moves out from the Sun quite rapidly — forming what is known as the solar wind. The wind therefore removes material from the Sun. As has been remarked previously, the amount lost is too small to affect the evolution of the Sun. Nevertheless, it is worth investigating whether other stars have extensive coronae which might dissipate material faster than on the Sun. Moreover, if stars can be found with coronae, this must mean, according to our theory, that they possess convective regions just below their surfaces. This can provide a check on our theories of stellar structure.

The solar corona is so tenuous that it is very difficult to detect even though we are close to the Sun. The problem of detecting stellar coronae is obviously even greater. However, the solar chromosphere can be detected quite easily. Similarly it has proved possible to detect signs of a chromosphere in the spectra of many stars. We can reasonably suppose that the presence of a chromosphere also implies the existence of a corona. It has been found that indications of a chromosphere are particularly evident in the spectra of the cooler stars, both giant and dwarf. It seems that the chromosphere becomes highly extended in bright, cool stars. We may interpret this as meaning that the atmospheric vibrations (often called turbulence) are more violent in the brighter stars.

One star that has been examined in great detail is α Herculis. This is actually a triple star. The two main components are a cool supergiant star and a giant with about the same surface temperature as the Sun. The latter has a close dwarf companion which is considerably hotter. The spectra of the giant star and its companion have been found to contain black absorption lines which cannot be assigned to either of them. It has been deduced that the lines are actually produced in the chromosphere of the supergiant star where it passes in front of the other two. But the supergiant and the giant are separated by a distance about 700 times greater than the distance of the Earth from the Sun. The chromosphere of the supergiant must be incredibly extended. The material in this chromosphere is moving outwards with much the same speed as in the solar chromosphere, but, because of the huge volume involved, there is a much greater loss of material from the supergiant than from the Sun. However, the supergiant is, in any case, evolving rapidly. Therefore this loss of material may still be unimportant: the star will have moved on to another stage in its evolution before the mass loss has much effect.

Neutrinos and the Sun

Chapter 3 contains diagrams (Fig. 24) of the proton-proton chain and the CNO cycle. If you examine these diagrams, you will see that one product of both sets of reactions is the neutrino. We must now look at this nuclear particle in a little more detail, for it plays an important role in stellar evolution.

The neutrino is a most unusual particle, for it has neither mass nor charge, and moves at the speed of light. As a result of the first two properties, neutrinos interact very little with other sub-atomic particles. In fact, the neutrinos produced by hydrogen burning at the centre of the Sun interact so little with the matter around them that most escape entirely from the Sun. Since they are moving at the speed of light, they make their exit only 2–3 seconds after they are produced. One consequence is that the Sun wastes some of its energy. As we have seen, the energy produced by nuclear reactions heats the interior of the Sun and prevents it from collapsing under its own gravitation. But the neutrinos carry away some of the energy produced: their rapid exit therefore means that this part of the energy is not used for heating. The amount removed from the Sun is small — only 3 per cent of the total energy produced. For stars at later stages in their development than the Sun, the amount of energy taken away by neutrinos can be much more significant.

Obviously, if we could only sample these escaping neutrinos, we could 'see' directly to the centre of the Sun, and so check that our theoretical picture of the solar interior is correct. At first, this seems impossible: if the neutrinos can escape from the Sun, how can they be captured on Earth? The answer is that their detection requires methods of extreme sensitivity, near to the limit of present-day techniques. A neutrino 'telescope' consists of a vast amount (say, a hundred thousand gallons) of some fluid containing the element chlorine. Tetrachloroethylene — a common cleaning fluid — is an obvious choice. It has been found that a chlorine atom can, very occasionally, capture a neutrino, turning itself into an atom of argon. The argon produced is, however, unusual, since it consists of a radioactive isotope. This radioactivity can be detected, so indicating that a neutrino has been captured.

Although the method is simple in principle, it is by no means easy to apply in practice. On the average, only one neutrino capture occurs per week: to obtain a significant number of events requires counting over a long period of time. Moreover, a flood of other nuclear particles reaches the Earth from space, and these can readily hide the effect that is being sought. (The experiment therefore must always be carried out at the bottom of a mine.) For these reasons, uncertainty over the results has continued for longer than with most experiments. Nevertheless, the overall outcome of the measurements so far is that the Sun clearly appears to be producing appreciably fewer neutrinos than the theoretical models predict.

If this result is accepted, our ideas of stellar evolution come under severe pressure; for it is generally accepted that we understand the Sun better than any other star. Is it really true then that theoreticians have got it all wrong? As it happens, there is one piece of observational evidence that suggests the opposite. It has been found that the Sun is pulsating: its surface is moving in and out in a regular way. Now, how a star pulsates depends on its internal structure. By measuring the Sun's pulsations we can therefore 'look' at its interior, just as we can by measuring the neutrinos emitted. The observational data on solar pulsation confirm the general accuracy of our present theoretical model of the Sun.

We are thus faced by a dilemma: should we trust the observations of neutrinos or of the pulsations? For the moment, we must wait for further evidence, hoping that the current theory will not eventually require excessive change.

CHAPTER 7

Red Giants and After

Exhaustion of the Central Hydrogen

We have seen in Chapter 5 that stars, whether upper, or lower, main-sequence, eventually burn all their central hydrogen reserves and move off the main sequence towards the right. We will now examine their subsequent progress. Reverting to our previous comparison of stars with plants or animals, we recall that a most vital function of living material is to produce energy. If a plant or animal cannot provide itself with a source of energy, it will very soon die. Similarly with a star — once it has started on its career it must continually produce energy: if it ceases to do so, it dies. When, therefore, the central hydrogen has been burnt, a star must develop some other source of energy. One is immediately to hand — contraction; after all, this was how the star started its career as an energy producer.

The centre of a post main-sequence star is composed mainly of helium with a slight admixture of heavier elements. This central region generates no energy. There is therefore nothing to prevent gravitational forces taking over once more, and causing the centre of the star gradually to contract. This has two consequences. Firstly the central region heats up; secondly the part of the star just outside this central core is pulled in closer to the centre. As a result, the temperature of the latter region rises. Formerly its temperature was slightly too low for hydrogen to ignite. Raising the temperature initiates hydrogen burning in a thin shell round the core. After a relatively short time all the hydrogen in the shell is converted into helium. What happens then, in effect, is that the new layer of helium is added to the

125

contracting core, and more unburnt material falls in from regions still farther out. The central regions of both upper and lower main-sequence stars during the earlier stages of their evolution from the main sequence therefore consist of a contracting core (composed mainly of helium) surrounded by a thin, hydrogen-burning shell. Owing to the increasing temperature, hydrogen burning now occurs by the CNO cycle for stars of all masses.

It might be expected that the outer envelope of the star, at this stage, would also be falling inwards, so that the star as a whole would be getting smaller. In fact, the exact opposite occurs. The contracting core is becoming continually hotter and brighter. This extra energy, acting on the lower-density envelope, expands it outwards. Indeed, we have the situation that, as the core grows denser and denser, so the envelope becomes more tenuous. In the H-R diagram the star continues to move away from the main sequence towards the right. This is because the outer layers of the star expand much more than central core contracts. The star, therefore, is growing larger.

The simplest theoretical model for a star at this stage in its development would appear to be straightforward. Most of its energy comes from the hydrogen-burning shell. We can suppose that the central helium core has much the same temperature throughout (since it is producing little energy). The temperature of this core is set by the surrounding shell. Finally, we must add on an exterior envelope. Unfortunately, this simple model has a major drawback. It was found some years ago that cores of the type described are only stable up to a certain limit. In fact, if the core weighs more than one-tenth of the total mass of the star, the model is unstable. This restriction on the relative size of the core is called the Schoenberg-Chandrasekhar limit. But our theory of main-sequence stars shows that both upper and lower main-sequence stars violate this condition. We can interpret this as meaning that the future evolution of these stars involves major internal readjustments. Our next task, obviously, is to describe what these changes are.

Further Evolution of a Lower Main-sequence Star

To discuss the evolutionary processes in detail, we must return to our former distinction between upper and lower main-sequence stars. This time we will take the lower main-sequence stars first. At this stage in their evolution the density of the central core is growing rapidly. We have based all our previous ideas of stellar evolution on the assumption that the whole of a star — even the central regions — can be thought of as a gas, and we have seen in Chapter 3 why this should be true. At the density with which we are now dealing — one hundred thousand times that of water — this assumption breaks down. The core material is being squashed to such an extent that it ceases to behave like a gas and starts to behave more like a molten metal. In fact, it begins to act rather like a large metal ball situated at the centre of the star; for example, it begins to conduct heat rather well. (This does not mean, of course, that the core has much metal in its composition. At this stage it consists mainly of helium; but helium under such extreme conditions has some properties rather similar to those of a metal.) Material in this 'squashed' state is said to be *degenerate*. The Schoenberg-Chandrasekhar limit was derived for stars with a gaseous helium core. It is not applicable to a degenerate core: the star has circumvented the limitation.

Degenerate material has another property that we normally associate with solids — it can withstand high pressures without an appreciable change in volume. When the core of a star becomes degenerate, it therefore ceases to contract. Our new model of the evolving star thus has a small, very dense, degenerate, central core composed mainly of helium. It is at the same temperature throughout: because degenerate material is a good conductor and any temperature differences are rapidly ironed out. Outside this is a partially degenerate shell of helium. (Partially degenerate material can be thought of as a mixture of degenerate and gaseous matter. As such, it is still capable of further contraction.) Outside this again there is a hydrogen-burning shell, and, finally, there is an extensive, but tenuous, envelope. As on the main sequence, the energy is carried through the outer parts of this envelope mainly by convection. The structure of the star at this point is shown in Fig. 45.

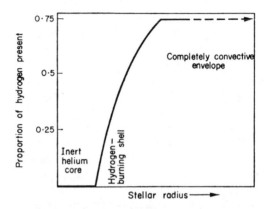

FIG. 45. Structure of a lower main-sequence star in the later stages of evolution from the main sequence.

As the evolution we are describing proceeds, the convective zone extends down farther: almost, but not quite, to the shell where hydrogen is being burnt. To get the scale right, remember that the shrunken core is now only a few times larger than the Earth, though the star as a whole has become several times larger than the Sun. This inward extension of the convection plays a major role in changing the direction of the star's evolutionary track in the H-R diagram. Formerly, it has been moving to a considerable extent from left to right, decreasing in surface temperature. Now it starts to move more in an upwards direction, becoming brighter. Its track therefore develops a noticeable elbow (Fig. 46).

Although the central part of the core is now static, changes may yet continue, for the outer regions can still contract. The hydrogen-burning shell is continually producing new helium. This is then added to the partially degenerate zone underneath, thereby slightly increasing its mass. The compaction towards the centre, besides increasing the mass of the core, also releases a certain amount of energy, and so raises its temperature.

The stars are now so swollen in size that they have reached the red giant region of the H-R diagram. Several of the stars in this region are known to be varying regularly in brightness. (These are the long-period variables mentioned in Chapter 2.) It is reasonably sure that the variations are due to the star pulsating in and out. The pulsations depend on conditions in the stellar envelope: they are believed to be due to some form of instability in the region where hydrogen and helium are becoming ionized. The

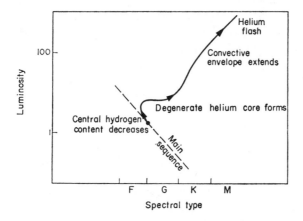

Fig. 46. The evolutionary track of a lower main-sequence star from the main sequence to the red giant region. Population I and Population II stars follow similar paths, but a Population II star moves into the red-giant region at a steeper angle in the H-R diagram than a Population I star.

period of the pulsation (the time from one light maximum to the next) depends on the detailed structure of the envelope. This is, of course, precisely what we have been trying to specify in our stellar models. We can therefore determine from our theoretical models what sort of period a star of this type should have, and then compare our answer with the actual observations. A comparison on these lines has been made for Population II red giants in globular clusters. (These stars were chosen because their ages, and, therefore, their exact states of development, were known.) The results are eminently satisfactory: both the calculations and the observations indicate a period of about a hundred days.

Helium-burning in Lower Main-sequence Stars

By this stage in its development, the temperature in the central core of our evolving star is several tens of millions of degrees. The helium nuclei at the centre are striking each other harder and harder, and more and more frequently. By analogy with the onset of hydrogen-burning we would expect that, sooner or later, helium nuclei would also ignite and start to provide energy. In fact, a comparison with the mechanics of hydrogen-burning suggests that helium should start to burn at much lower densities and temperatures than we have now reached. The answer, as we have seen in Chapter 3, is that our analogy is inexact: the interaction of two helium nuclei differs fundamentally from the interaction of two hydrogen nuclei. The hydrogen nuclei combine to form a stable deuterium nucleus, but the helium nuclei combine to form a beryllium nucleus, containing eight particles, which is extremely unstable and breaks down almost immediately into the two original helium nuclei. In order to convert helium into another substance, it is necessary that three nuclei should strike together simultaneously. Since the probability of three helium nuclei hitting each other at exactly the same time is very small indeed, the collisional rate needs to be very high before a significant number of reactions will occur. This is why the core of our post main-sequence star must become very hot and very dense if helium is to ignite. Added to this are two further factors. Firstly, the high conductivity of the core allows heat to leak away rapidly. Secondly, a considerable number of neutrinos are likely to be produced at such high densities. As we saw when discussing the Sun, energy that would otherwise have helped heat the core is thereby lost to space. The triple-alpha process becomes possible when the density corresponds to that of degenerate matter and the temperature approaches a hundred million degrees.

When constructing stellar models, it is vital to confirm always that they are stable. If due consideration is not given to this point, it would be quite possible to construct an excellent theoretical stellar model which, if it ever came to exist in nature, would immediately fall to pieces. The difference between a stable and an unstable star is rather like the difference between a tricycle and a bicycle. The former remains upright (i.e. stable) unless it is actually tipped over, the latter will only remain upright if it is supported. We know from observation that the huge majority of stars are stable: they do not change very much over reasonable periods of time. We must therefore ensure that our stellar models are similarly stable, or, at least, that any instability they contain disappears rapidly. As it happens, the stability requirement is generally satisfied for a sphere of ionized gas (corresponding, for example, to our picture of a main-sequence star). Suppose we consider what would happen to the Sun if we reduced the amount of energy being generated in its interior. We have seen that it is this energy which keeps the Sun at its present size. If the energy is cut down, the Sun will start to contract. The contraction produces energy which heats up the solar interior. This, in turn, increases the rate at which hydrogen nuclei interact in the central core. An increased amount of nuclear energy is therefore produced, and very rapidly the Sun returns to its initial state. Similarly, if we increased the amount of energy produced in the core, the Sun would expand. This would cool it; the rate of nuclear reaction would decrease, and, once again, we would be back where we started. In other words, the Sun has a stable balance of energy.

Our evolving star, however, cannot be fitted into the same picture. The central core material is now not gaseous, but degenerate. The solar energy balance depends on the Sun's ability to compensate for energy variations by contracting or expanding.

Degenerate material, being more like a solid than a gas, cannot alter its volume in a similar way. Consider then what happens when the temperature in the degenerate material rises high enough for the triple-alpha process to start. Energy is immediately released, and the temperature increases still further. But, because the material is degenerate, this heating does not cause an expansion as it would do in an ordinary star. Instead the degenerate material heats up rapidly. This increases the rate at which the helium nuclei interact (since the collisional rate is increased). Much more energy is generated, and the temperature shoots up. Very quickly, a run-away process develops; the whole situation is completely unstable and a sudden burst of energy is produced in the core.

Now, however, a new factor appears. Whether material is degenerate or not, depends on its resistance to compression. It takes less work to squash cold material together than hot, because hot material is trying to expand with greater energy. This means that a pressure which is great enough to make material at a certain temperature degenerate, may not be sufficient to keep it degenerate at a higher temperature. It turns out, in fact, that the burst of energy developed when helium ignites is sufficient to convert the degenerate 'solid' material at the centre of the star back into an ionized gas. This gas, when it appears, finds itself at a much higher temperature and density than it should be, so it expands very rapidly to try and reach more acceptable conditions (i.e. it tries to return to a stable situation). The whole process from the ignition of helium to the expansion of the core is called the *helium flash*. As the name implies, the helium flash is a rapid event. For a low-mass star, however, the time taken from leaving the main sequence to reaching the helium flash point can be long. A star of mass $\frac{3}{4} M_\odot$ (typical for a Population II star in a globular cluster) will take about 2000 million years for the transition: that is some 15 per cent of the time it remained on the main sequence.

At this stage in its evolution, a star similar to the Sun will have almost 50 per cent of its total mass in the helium core, and it will be over a hundred times brighter than the Sun. Its surface temperature, however, will be mainly controlled by the huge, tenuous, outer envelope, so it will appear quite cool — a few thousand degrees only. This is, of course, a description of a typical red giant. We find, in fact, that most lower main-sequence stars over a wide range of masses experience the helium flash when they have reached about the same brightness. To put it another way, there is a restricted region in the H-R diagram, coinciding with the positions of the red giants, where lower main-sequence stars may be expected to undergo a sudden change in their evolutionary development.

The Effect of the Helium Flash

The helium flash is obviously a most unstable process. Indeed, the nearest terrestrial comparison would be a bomb explosion. The most obvious effect of a bomb explosion is to shatter anything in the near vicinity into fragments. We must obviously decide first whether our evolving star may not be similarly fragmented. It is quite difficult to calculate accurately the consequences of the helium flash. Unlike the evolutionary stages we have examined so far, where change is slow, the star now alters significantly within a few seconds. Indeed, computing the stellar models takes much longer than the actual evolution that is being modelled.

As the flash occurs, the star 'pops' outwards, and then pulsates in and out for a while. Some material from the surface regions may be lost to space, but the outer envelope effectively tamps down the core and prevents a major catastrophe. In fact, the force of the explosion even seems insufficient to cause any appreciable amount of mixing in the stellar interior. The end result of the flash is therefore a star burning helium to carbon, and later to oxygen, in its core, and burning hydrogen in a surrounding shell. Here, for the first time, we encounter a star with two simultaneously burning, but separate, nuclear energy sources. One consequence of this complication is that it becomes more difficult to predict where a given star will lie in the H-R diagram, and how it will change position in the diagram as it evolves further. One rule-of-thumb can usually be applied: that the type of motion to be expected in a star — whether expansion or contraction — will change on either side of a nuclear-burning shell. For example, during a star's evolution we have seen that the core contracts *en route* to the red-giant branch. We would therefore expect that the envelope should expand, since it is on the other side of a nuclear-burning shell (burning hydrogen in this case). This, of course, is precisely what we observe.

Basically, a star, after the helium flash, settles itself initially somewhere on the horizontal branch, which we saw as a typical feature of globular cluster H-R diagrams. The exact position depends on the mass and chemical composition of the star. The lower the mass and the smaller the proportion of elements heavier than helium, the bluer the star will be. Hence, the further to the left it will lie on the horizontal branch. This at once clears up one of the obvious differences between the H-R diagrams of galactic and globular clusters: whereas the latter have horizontal branches, the former do not. The reason is evidently that the evolving stars in galactic clusters are both more massive and contain more heavier elements than their equivalents in globular clusters. As a result, they remain at the red end of the horizontal branch and cannot readily be distinguished from the red giant branch.

The direction of evolution from this initial position depends primarily on the relative contributions of the H-burning shell and the He-burning core to the total luminosity of the star. The larger the contribution of the former, the more likely it is that the star will evolve from red to blue. Conversely, a relatively greater contribution from the core is likely to lead to evolution from blue to red. We can tentatively relate the existence of these two possible directions of evolution to some of the observational evidence concerning the horizontal branch. Of course, the rate of evolution is very slow in terms of human lifetimes. It is therefore not possible to observe directly which way the stars are evolving. However, as we have seen, the horizontal branch contains a region of instability where the RR Lyrae stars lie. The instability here is superficial: it affects the outer layers of the stars, rather than the deep interiors. Hence, the evolution of the star is not affected. But it is found that there are detailed differences in the way RR Lyrae stars in different globular clusters pulsate. It has proved possible to relate these differences to the direction of evolution through the instability strip. To this extent, one can point to some backing for the evolutionary picture presented here.

As the helium burns in its core, a star follows a fairly contorted path in the H-R diagram. It moves backwards and forwards from red to blue, becoming constantly brighter, and so reaching a new 'suprahorizontal' branch. Ultimately, all the helium in the core is burnt up; as the hydrogen was before it.

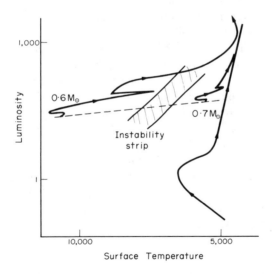

FIG. 47. Evolution up to, and after, the helium flash for a 0·7M☉ star, and after the helium flash for a 0·6M☉ star.

For nuclear burning beyond hydrogen, the core is always convective, regardless of the mass of the star. Helium is therefore totally consumed throughout the entire central region of the star.

The star now contains two shell-burning sources. The first of these is the old hydrogen-burning shell at the base of the envelope. The other is a new helium-burning shell round an inert oxygen-carbon core. The star shifts from blue to red and becomes markedly brighter, evolving upwards in the H-R diagram parallel to the original red-giant branch. (It is said to have reached the 'asymptotic' branch.) The rate of evolution is speeding up. A typical lower main-sequence star will move from the horizontal branch to the asymptotic branch in considerably less than a hundred million years.

It is at this stage that forecasting subsequent evolution becomes particularly difficult. As the star climbs up the asymptotic branch, there is a marked tendency for it to become unstable and variable in energy output. The pulses are spaced at very long intervals, unlike those in the RR Lyrae phase, but they are of much more importance. Whereas the RR Lyrae instability affected only the envelope, this new instability is deep-seated and can affect the nuclear-burning regions. In consequence, three things may happen.

There is a strong likelihood that some mixing will take place. In view of the complex burning situation inside the star, such mixing can lead to the formation of a wide range of heavier elements, and these will be brought to the surface where their existence can be detected. It is possible that this process has actually been detected. A unique variable star, FG Sagittae, has been observed to brighten during the present century. At the same time, its spectrum has shown the appearance at its surface of a mixture of heavier elements (such as barium and strontium).

Mixing can also lead to excursions in the H-R diagram: the star loops out towards the blue and then returns towards the red-giant branch again. One interesting point about this development is that it can lead the star to cross again the instability strip in the H-R diagram. Being more luminous at this stage, it would not cross the RR Lyrae region, but higher up, where the Cepheids are found.

The most important consequence of instability is, however, the possibility of mass loss occurring. It is difficult to be sure how much material may escape, but in view of the basic nature of the instability, it could be an appreciable fraction of the total mass. In fact, it seems possible that some stars at this stage can blow off almost all their entire envelope. The result of such an explosion would be something very similar to the observed planetary nebula — small, hot central stars surrounded by an expanding shell of gas. The fate of such stars will be taken up in the next chapter.

Further Evolution of an Upper Main-sequence Star

We have seen that the Schoenberg-Chandrasekhar criterion sets an upper limit to the size of the helium core, if the star is to be stable after it leaves the main sequence. Initially, when an

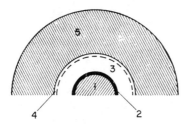

FIG. 48. Structure of an upper main-sequence star in the later stages of its evolution from the main sequence.
1 – inert carbon-oxygen core; 2 – helium-burning shell; 3 – inert helium layer; 4 – hydrogen-burning shell (possibly inactive at this stage); 5 – envelope still rich in hydrogen.

upper main-sequence star begins burning hydrogen in a shell round the core, this limit is not violated. To a large extent, the core can support the weight of material above it, and evolutionary change of position in the H-R diagram is limited. However, as more helium is added to the core from the burnt material in the shell, a stage is reached where the criterion is violated. At this point the core begins to contract rapidly, and the star moves fairly rapidly to the right in the H-R diagram (in the same manner as lower main-sequence stars).

The important difference between the upper and lower main-sequence star derives from the different central conditions. The upper main-sequence star starts off with a higher central temperature and, as we have seen, also with a lower central density than the lower main-sequence star. Consequently, helium ignition occurs before the material in the core becomes degenerate. In the H-R diagram there is therefore only limited change in luminosity, so that, unlike the lower main-sequence star, little vertical motion in the H-R diagram occurs. The upper and lower main sequence channel together in the red-giant region: there is little difference in position in the H-R diagram for stars of (say) $M_\odot$ and $3M_\odot$ as they approach the helium ignition point. The same is not true, of course, for much more massive stars. A star of $15M_\odot$, for example, which evolves horizontally, moves towards the region of red supergiants, rather than towards the red giants. In fact, for such a star the core conditions for helium burning are reached fairly soon after it leaves the main sequence, whilst it is still in the blue region. But later stages of evolution do take it into the red region. (After helium-burning in the core, a massive star moves on to the same double shell-burning stage that lower main-sequence stars experience.)

One important consequence of helium ignition occurring in non-degenerate material is that no instability problems arise. Any internal change is slow. Hence, at this point, evolutionary stellar models can be computed more readily for the upper main sequence than for the lower main sequence.

Mass Loss and the Evolution of Lower-mass Stars

We have seen that lower main-sequence stars are likely to lose material as they evolve to the helium shell-burning stage. We shall see shortly that upper main-sequence stars are also likely to lose material in the later stages of their evolution. Mass is one of the basic factors in determining stellar evolution, so the eventual mode of decay and death of a star will depend to a great extent on how much mass it loses during its lifetime. This, in turn, depends both on the rate at which mass is lost and the period of time over which it is lost. For example, massive stars evolve rapidly; so, even if they lose material quickly, they may still have time to run through their evolutionary careers with little effect.

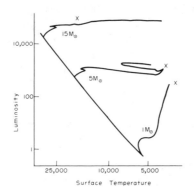

FIG. 49. The evolutionary track of upper main-sequence stars after leaving the main sequence. (X indicates the beginning of helium burning for each mass.)

Extensive mass loss can fundamentally alter stellar evolution by changing the energy available. Because a nuclear fuel is available in the core of a star, it does not necessarily mean that it can actually be utilized by the star to produce energy. This depends on the appropriate temperature and density for ignition being reached in the core, which is determined by the stellar mass. It is, in fact, possible to define mass limits for stars, beyond which they cannot progress to the next nuclear-burning stage. Roughly speaking, a star of mass less than $\frac{1}{2}M_\odot$ cannot ignite helium; a star of mass less than $\frac{3}{4}M_\odot$ cannot ignite carbon; and so on. Consequently, large-scale mass loss at any stage in a star's evolution can prevent it from progressing to later evolutionary stages. What happens then can be seen from an examination of how stars at the bottom of the main sequence evolve.

Suppose we consider a star of mass $\frac{1}{2}M_\odot$, or less, on the main sequence. Its initial evolution is much the same as for any other main-sequence star: a helium core develops which becomes partially degenerate as the star moves away from the main sequence. At this point the evolution diverges. A normal main-sequence star builds up its central temperature by continually adding new helium to the core from the hydrogen-burning zone. A very low-mass star, however, only has a limited amount of material outside its core. For a star of mass $\frac{1}{2}M_\odot$, this amount is

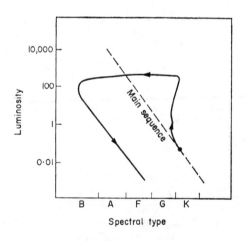

FIG. 50. Post main-sequence evolution of low-mass stars.

insufficient to create a central temperature which will ignite helium. The helium flash can never occur for such a star. Instead, it consumes as much of its hydrogen as it can, then it grows smaller and hotter and slides across the diagram towards the region of the white dwarfs. (Actually, a star with a mass as low as this evolves very slowly: it may take longer than the present age of our Galaxy to reach the white dwarf region.)

The Evolution of Double Stars

Our foregoing discussion has, as usual, been concerned only with single stars. We must now consider whether there may be differences in the evolution of double stars. We can surely feel certain that in a wide double star each component will evolve separately as if it were a single star. We will therefore consider only very close doubles.

Suppose (as is often the case) that one component is more massive than the other. On the main sequence, the former will be the brighter star, and is therefore called the *primary*. The lower-luminosity component is correspondingly referred to as the *secondary*. The more massive star — it may be on the upper

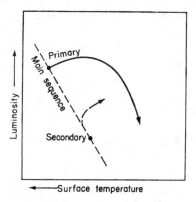

FIG. 51. Schematic diagram of the initial evolution of the primary and secondary components in a close double star.

main sequence — will consume its central hydrogen more quickly and will evolve away from the main sequence first. As soon as a helium core develops, it will start rapidly to expand. But now it encounters an obstacle: it has a nearby companion. As the envelope of the primary expands it begins to engulf the secondary. The gas in this envelope is thus closer to the surface of the secondary star than it is to the core of its own star. This means that the companion star can exert a greater gravitational pull on the gas than its original owner can. It therefore immediately starts to absorb the material and adds it to its own surface. As a result, the primary loses mass, and the secondary grows.

If the evolutionary path of the primary star is plotted in the H-R diagram, we find that this loss of material makes the primary fainter. Instead of moving across towards the red giant, or red supergiant, region, the star's path turns downwards and it approaches the region of the subgiants. This, as it happens, is precisely what the observations demand. It is a long-recognized peculiarity of close double stars that one component is often a subgiant, yet subgiants are quite rare among single stars.

The further development of close double stars can only as yet be guessed. It is possible that the transfer of material goes on until the primary and the secondary change their roles — the primary becoming less massive and the secondary more massive. Since the rate of evolution depends on the mass, this may mean that the secondary now begins to evolve more rapidly than the primary. If so, retribution is close at hand. The secondary quickly finishes off its central hydrogen reserves and expands. This causes it to overlap the original primary star which immediately reclaims its lost material. These mutual interchanges cannot, however, go on for ever. For one thing, the calculations show that not all the material pulled off the primary will actually be acquired by the secondary. Some of it may be affected in such a way that it either orbits round the double star or else disperses altogether and evaporates into space. Sooner or later this overall loss of mass will inhibit the activity of both stars.

We can see qualitatively that close double stars must be limited in the extent to which they can evolve across the H-R diagram from left to right. Indeed, the type of evolution outlined above will inevitably force them back to the left-hand side of the diagram (though this may only occur after they have suffered considerable loss of mass). Most of the close double stars that have been studied in detail belong to Population I: no bright, close binaries have been found in globular clusters. However, a small number of stars in globular clusters are observed to lie on the main sequence in the H-R diagram, but above the normal turn-off point for the cluster. These so-called 'blue stragglers' may actually be close double stars whose evolution has been affected by interaction between the components.

We have seen in Chapter 5 that Wolf-Rayet stars which are members of associations may be mixed stars immediately post main sequence. Another type of Wolf-Rayet star is found which is not connected with associations. It differs from the first type in having an excessive amount of carbon in its atmosphere rather than an excessive amount of nitrogen. We can explain this second type of Wolf-Rayet star by supposing that it has passed through the helium-burning state, and the carbon produced has somehow been mixed to the surface. Thus both types of Wolf-Rayet star may show the effects of mixing, but at different stages in their evolutionary careers.

Nuclear Reactions after Helium burning

We now turn to the later stages of a star's evolution, after helium burning. It must be remembered that ideas concerning these later developments are little more than intelligent guesses. Relatively few models have been calculated in detail, and even these have had to be based on some rather debatable assumptions. Only one thing is completely certain and that is that a star must sooner or later use up its energy reserves and complete its evolution.

We must consider first what nuclear reactions can occur after the helium in the core of a star has burnt to carbon. They will depend, of course, on the density, temperature and chemical composition at the centre. This last point is, perhaps, the most important for, as we have seen, it can be altered by mixing. In an unmixed star each layer in towards the centre will be more deficient in the lighter elements: the outermost layer has plenty of hydrogen, the next has no hydrogen but plenty of helium, the next has neither hydrogen nor helium but plenty of carbon, and so on. In a completely mixed star these distinctions vanish. Hydrogen is burnt until it has disappeared completely from the whole star; only then does the helium ignite. If there is an intermediate amount of mixing, the relative proportions of the nuclear isotopes produced will vary with the degree of mixing and the evolutionary stage at which it takes place. We will discuss the general picture of nuclear burning on the assumption that mixing can be ignored.

The first point to notice is that the production of a carbon nucleus from three helium nuclei is not the endpoint of helium burning. A helium nucleus may interact further with one of the carbon nuclei and build it up into oxygen (containing sixteen sub-atomic particles). When the temperature reaches six hundred million degrees (and the density is some ten thousand times greater than that of water) the carbon nuclei start to react with each other. This produces more oxygen, neon and magnesium together with some sodium. It also releases a flood of nuclear particles — such as protons and neutrons — which react with the various nuclei present to form other nuclei of intermediate weight. As the temperature continues rising past a thousand million degrees, new reactions come into play such as, for example, the interaction of oxygen nuclei with each other. Somewhat heavier elements are now built up, including silicon, sulphur and phosphorus. A further increase of temperature towards three thousand million degrees induces an incredible profusion of reactions. Nuclei of a large number of elements are rapidly created up to, and including, iron, which contains 56 sub-atomic particles.

Ultimate Evolution of Main-sequence Stars

We now turn to the later stages of a star's evolution, after helium burning. We have noted that there are two major uncertainties in stellar evolution from this point onwards — the extent to which stars lose mass and the amount of energy carried away by neutrinos. Our lack of knowledge makes it difficult to assess both the nature and duration of the final evolutionary stages. One thing can be said with confidence: the stars that reach these stages must mainly have originated on the upper main sequence. Indeed, if some of the estimates of mass loss are correct (for example, that a $3\frac{1}{2}M_\odot$ star is reduced to less than $1\frac{1}{2}M_\odot$ in its later stages), only the upper end of the original main sequence need be considered.

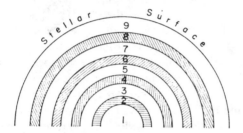

FIG. 52. The possible structure of a highly evolved star. 1 – mainly iron core; 2 – layer mainly of silicon; 3 – layer mainly of magnesium and silicon; 4 – layer mainly of magnesium, silicon and sulphur; 5 – layer mainly of oxygen and magnesium; 6 – layer mainly of oxygen, neon and magnesium; 7 – layer mainly of carbon and oxygen; 8 – layer mainly of helium; 9 – layer mainly of hydrogen. Between each layer there is a shell burning the appropriate fuel, e.g. between layers 8 and 9 there is a shell where hydrogen is being burnt to helium.

The critical question is whether, or not, carbon burning begins in material that is degenerate. It seems likely that the 'carbon flash' in degenerate material will be a good deal more violent than the 'helium flash'. Whereas the latter does not lead to a major rearrangement of the star's structure, the former probably will. Unfortunately, the mass at which degeneracy occurs cannot be easily defined, not least because of the uncertainty concerning energy losses via neutrino emission. In terms of stellar masses on the main sequence, it seems possible that any star with an original mass of less than $5M_\odot$ will develop a degenerate carbon core. On the other hand, any star with an original mass of more than $10M_\odot$ is likely to ignite carbon (and oxygen) before the core becomes degenerate. Where the changeover occurs in the intermediate mass range between $5M_\odot$ and $10M_\odot$ remains uncertain.

The effect of neutrino emission, since it removes energy, is to speed up the rate at which stars evolve. For example, if no neutrinos were lost to space, the carbon-burning phase of a typical upper main-sequence star would be extended by a factor of at least a hundred. But, so long as the material in the star remains non-degenerate, the direction of evolution remains unchanged. The successive evolutionary stages are therefore reasonably predictable for a sufficiently massive star. At each stage, once the nuclei in the core have completed all the reactions of which they are capable, burning ceases, and the core contracts. As soon as the central temperature reaches a high enough value for the next set of reactions to start, a new convective core develops. Meanwhile, the old set of reactions will have been pushed outwards, and will be occurring in a shell round the core. (The amount of energy produced by any given shell may, however, change markedly with time.) We can, in principle, follow this type of process through to the point where iron is being formed at the centre. By this time there may be as many as nine different layers in the evolving star. At this stage, the direction of evolution does change radically, as we will see in the next chapter.

CHAPTER 8

Old Age and Death

WHEN a star begins to produce iron at its centre, it has reached a turning point in its evolutionary career. We have seen that successive nuclear reactions produce less and less energy. (This plays an important part in speeding up the later stages of stellar evolution.) Thus burning hydrogen to helium gives ten times as much energy as burning the equivalent amount of helium to carbon. Alternatively, one can say that the mass which disappears when four protons are converted into one alpha-particle is greater than the mass lost when three alpha-particles are converted into a carbon nucleus. So it continues as far as iron. But when we try to add nuclear particles to iron, in order to convert it into a heavier element, we find that instead of getting energy from the process we actually have to provide energy. Iron is therefore the end of the road for nuclear burning: we can get no energy from building up heavier nuclei. How, then, can our evolving stars cope with this crisis?

The End of Upper Main-sequence Stars

What happens to the most massive stars seems clear. At temperatures of a thousand million degrees, or more, the neutrinos begin to remove a high proportion of the energy generated very quickly. The core therefore contracts rapidly to try and keep up with the energy loss. Changes now become evident in the structure of the star not over periods of a million years, but over a few years, and the central temperature may shoot up to several thousand million degrees. At such a temperature the particles are moving extremely fast, but, because the density is high, they are also striking each other very frequently. Eventually, the iron nuclei present in the core can withstand the impacts no longer and disintegrate. The resulting fragments are either neutrons, or else the very stable helium nuclei which are able to survive bombardment even under these extreme conditions.

This breakdown of iron nuclei has very serious consequences for the star. We have seen that the later stages of stellar evolution have derived their energy by the slow build up of helium nuclei into iron. Now the reverse process has suddenly occurred. If building up helium to iron releases a certain amount of energy, then breaking iron down to helium must absorb that same amount. This means that the energy produced by the star over a million years or more is required back almost instantaneously. We might compare the star with a human being who has accumulated a large overdraft at his bank over a long period of time. Suddenly, the bank manager demands immediate repayment. A human being in these circumstances would probably go bankrupt; so, in a much more spectacular way, does a star. The star has only one asset left: its ability to contract and convert its gravitational energy into heat. Now it must use this asset to replace a huge energy deficit almost instantaneously. The resulting contraction is no longer a slow, gentle process, but a sudden collapse — like a balloon that has been pricked.

But the star overdoes things. Like some human bankrupts on similar occasions it seems bent on suicide. The collapse of the central layers certainly repays the energy debt due to the fragmentation of iron into helium. But the outer layers of the star, deprived of their support, also collapse. These layers are at considerably lower temperatures, and still contain nuclear fuel available for burning. (The outermost layers of the star may still, indeed, retain their original composition.) This material is now suddenly subjected to a very high temperature. It has been pointed out before that the speed at which nuclear reactions occur depends very sensitively on the temperature. The collapsed outer material therefore becomes the seat of a frenzied activity as nuclear-burning processes, which normally take thousands, or millions, of years, go to completion within a few seconds. The situation is obviously violently unstable: we can best describe it as a vast nuclear explosion, for the amount of energy generated is more than enough to halt the collapse and to eject all the stellar material into space.

The End of Less Massive Stars

It is not only the most massive stars that finish their lives in a crisis of instability. Stars of intermediate mass (say, the lower part of the upper main sequence) also run into difficulties as they approach their ends. The problem here is that they ignite carbon when it is in a degenerate state. This is likely to lead to a run-away reaction, with energy losses mounting rapidly, and to the body of the star, itself, being dissipated into space. Similar, though less violent, instability characterises stars on the lower main sequence. As we have seen, helium-burning stars eventually climb the asymptotic branch, where they show a marked tendency to become unstable. Here the result may again be massive ejection of material from the star.

Clearly, any stellar object that remains after these episodes of violent mass loss still faces the problem of reaching some kind of equilibrium. If it can contract, it will do so; but most such remnants will already be of high density. Consequently, the final stage of all these stars will consist predominantly of degenerate matter. The query is whether a quiescent ball of such matter, devoid of energy sources, can be in equilibrium. To put it in other words — Has degenerate material sufficient strength to resist for all time the gravitational pull inwards that is trying to compress the star still further? The answer must depend on the mass of the star. If, after mass loss, only a small stellar core is left, then the strength of its degenerate matter can certainly counteract the gravitational forces. But, for more massive stars, the resistance of the degenerate material to compression increases only slightly, whilst the gravitational pull inwards increases rapidly. Hence, there is a limiting mass above which degenerate material can no longer hold out against the weight of the star. This limiting mass is known as the *Chandrasekhar limit*, and cannot be higher than about $1\frac{1}{2}M_\odot$ (It may be appreciably lower, depending on the chemical composition of the star.) We will take up later in this chapter the question of what happens to stars with masses exceeding this limit. We note here that all lower main-sequence stars and (because of mass loss) many upper main-sequence stars can satisfy this criterion for ultimate stability.

The only stars that may avoid some period of instability in their final development are those right at the bottom of the main sequence. As we have seen before, these stars have fairly high central densities and entirely mixed interiors on the main sequence. Moreover, they are not massive enough to ignite helium. Such stars may be able to reach a stable degenerate configuration with little fuss. One remaining uncertainty is the length of time they take to evolve to this final stage. The evolutionary timescale expands rapidly as one moves down the main sequence. It might be supposed, therefore, that low-mass stars would take much of the lifetime of our Galaxy to reach their end state. It is possible, however, that these stars may evolve somewhat more rapidly than we have supposed. For example, the planet Jupiter can be thought of as a very low-mass star which never reached a sufficiently high internal temperature to ignite its hydrogen. Yet it has obviously managed to condense well within the lifespan of our Galaxy.

Novae and Planetary Nebulae

Rapid ejection of mass is an explosive event, which necessarily produces a strong burst of radiation. In observational terms, this means that the star undergoing such mass loss flares up in brightness, not only as seen visually, but at other wavelengths, too. In looking for stars that are reaching the end of their evolutionary careers, we therefore look for violent outbursts. At the same time, our ideas on evolution tell us that such stars will often be blue in colour, and of fairly high density. With these points in mind, we can try to identify the final stages of the less massive stars.

Novae explosions, as we have seen, throw off only a small amount of material, and are therefore by no means catastrophic (unlike a supernova explosion). On the other hand, since the explosions probably recur at intervals, the nova mechanism may eventually get rid of an appreciable quantity of stellar material. As novae are members of close binary systems, we can determine a good deal about their characteristics. We find, in fact, just as we might hope, that novae are small, dense stars which fit in well with our conception of a star evolving towards the white-dwarf region. However, the close binary nature of novae also precludes us from using them to explain the final evolution of single stars (with which we have mainly been concerned in previous chapters).

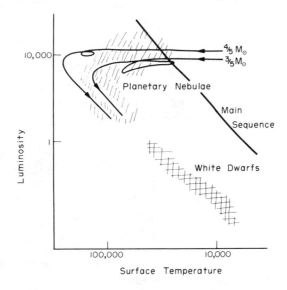

FIG. 53. Theoretical evolution of planetary nebulae to white dwarfs compared with observational data.

promising from this point of view. They appear to be single stars, in general, and, although this prevents us from determining as much about them as we can for novae, nevertheless, it seems reasonably sure that they are smallish, dense stars of the type we are wanting. The major problem with planetary nebulae is that, although many are known, and although there is no doubt that the gaseous shells are expanding outwards, no explosion producing such a nebula has ever been observed. We can, however, argue round this difficulty in the following way. We know that the frequency with which a nova explodes depends on the size of the explosion: the larger the explosion, the more infrequently it occurs. Now the shell of a planetary nebula contains far more material than the shell thrown off by a nova. We can therefore suppose that the explosion forming a planetary nebula is very large, and so occurs very rarely.

White Dwarfs

We now have some idea of where, in observational terms, the less massive stars begin their descent to ultimate oblivion. But whereabouts in the H-R diagram are the quiescent spheres of degenerate matter, that we expect to appear from the explosions? There is one group of stars that precisely fits the bill — the white dwarfs.

White dwarf stars are rather misnamed, as we remarked in Chapter 2. Their surface temperatures range from 40,000°C down to 4000°C, so that although several of the brighter ones are, indeed, white in colour, many fainter members of this group are yellow, or even red. When plotted in the H-R diagram, white dwarfs tend to lie along a line approximately parallel to the main sequence, although, of course, much below it. Their position in the diagram indicates that they must be very small — not much larger than the Earth. This remarkably small size is tied up with the distinguishing feature of white dwarfs: their density. A few white dwarfs are members of double stars. (In fact, the first member of this group to be discovered was the white-dwarf companion of the Dog Star.) It has therefore been possible to measure the masses of a few white dwarfs, the average value coming to about $\frac{1}{2}M_\odot$. If we combine this with their size, we find that their mean density may be as much as a million times that of water. This sort of density implies that white dwarfs are degenerate throughout virtually the whole of their volume; only a thin surface layer can act like a normal gas. (It is, however, this surface layer which produces the characteristically peculiar spectra of white-dwarf stars.)

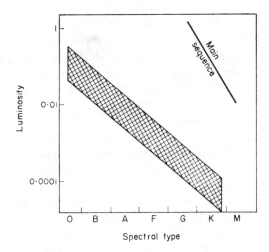

FIG. 54. Schematic diagram showing the general area of the Hertzsprung-Russell diagram occupied by white dwarfs.

Since degenerate material is a good conductor, white dwarfs will be at virtually the same temperature throughout. This contrasts with normal stars where the temperature decreases very rapidly outwards from the centre. We have compared white dwarfs previously with metal spheres suspended in space. There is one modification we must make to this picture: the thin atmosphere forms a layer of insulation round the degenerate interior so that the heat trickles out only very slowly. (It acts rather like the material round a hot-water tank by preventing too rapid a loss of heat.) A white dwarf therefore loses heat quite slowly, becoming gradually cooler and fainter. This is the reason why there is a 'white-dwarf sequence' parallel to the main sequence: it represents the path in the H-R diagram along which white dwarfs lose their heat. On this interpretation, the very faint, red 'white dwarfs' are the ones that have been cooling down for a long period of time. The brighter white dwarfs — those still white in colour — are relatively recent arrivals.

We have seen that degenerate material is subject to Chandrasekhar's limit. Since virtually the whole of a white dwarf is degenerate, Chandrasekhar's limit now defines an upper bound not simply to the mass of the core, but to the mass of the whole star. We can state categorically that no white-dwarf star may be more than $1\frac{1}{2}M_\odot$. The few observations of white-dwarf masses available fall satisfactorily below this limit.

White dwarfs are continually losing heat, and have no energy sources with which to counterbalance this loss. Their final fate is therefore inevitable. They must ultimately dissipate all their internal energy into space and become dead bodies at a uniform temperature close to absolute zero. Such 'stars' have been called 'black dwarfs': they cannot be observed directly, but might be detected indirectly by their gravitational effects. However, the time required to reach this final phase is very long, since white dwarfs cool off only slowly. The process may take longer than the present lifetime of our Galaxy for most stars; in which case few black dwarfs will yet have formed.

Supernovae

We have seen in Chapter 2 that major explosions in stars — the supernovae — are well known, if not very frequent, happenings. However, there are two types of supernova: the Type I supernovae related to the old Population II stars and the Type II supernovae related to the young Population I stars. These two groups have characteristics that reflect not only their chemical and age differences, but probably also the different evolutionary stages reached by the pre-supernovae in the two populations. Thus not only is the material ejected by Type I supernovae deficient in hydrogen as compared with Type II supernovae (something we would expect if the former come from the older population), but also the Type I supernovae throw out a much smaller quantity of material than the Type II. This implies that Population I supernovae are massive stars, whereas Population II supernovae are of a lower mass. The cause of this difference seems obvious. On the one hand, there are no massive Population II stars left; on the other, the lower-mass Population I stars have not yet reached the end of their evolutionary careers. So we have violent explosions occurring in stars at two differing stages. Our next step must be to identify these stages.

The explanation of the massive Population I supernovae presents few problems. We have already found that such stars reach a crisis in their development after the appearance of iron in their cores. The resulting explosion certainly produces the amount of energy required for a Type II supernova. But what is the origin of Type I supernovae? It seems clear from theoretical calculations that the other place where instability can occur is nuclear ignition in degenerate material. We have seen that this may take place at more than one stage in a star's development. However, the helium flash does not generate the level of instability required for a supernova explosion. Instead, we must turn to those stars that develop degenerate carbon-oxygen cores. Here, though there are a number of uncertainties in the theory, violent ejection of material does seem possible, Hence, we can tentatively identify this stage of evolution as providing a source of lower-mass supernovae.

Neutron Stars

What remains after a supernova explosion? Obviously, there are vast, expanding clouds of gas, which form the easily observed supernova remnants. But can any stellar remnant be found? In at least some cases, the answer is certainly 'yes'. We can take as our example the Crab nebula — by far the best observed supernova remnant.

It was found at the end of the 1960's that a small, peculiar star near the centre of the Crab nebula emitted pulses of radio noise thirty times every second. (It appeared subsequently that the star emitted similar pulses of radiation at a variety of other wavelengths, too.) The star was therefore identified as a member of the then newly discovered group of 'pulsars'. The observational data, especially the rapid rate of pulsing, imposed stringent limits on the nature of the central star. The pulses could only be explained theoretically as resulting from the rotation of the star; but no normal star could spin at such a rate. If the Sun, for example, rotated at this speed, it would very soon shed most of its material into space. The gravitational pull would be nowhere near strong enough to prevent the matter from flying off from the equator. Only a very small, dense star — much smaller and denser than a white dwarf — could avoid this fate. In fact, the Sun would have to be packed down into a sphere only a few miles across to be completely safe.

How can we explain pulsars in theoretical terms? We must first go back to a related question that has so far been left unanswered. What happens to a dying star, if its mass exceeds the Chandrasekhar limit, so that it cannot become a white dwarf? The problem for such a star is that the pressure exerted by degenerate material is insufficient to resist gravity, It must therefore fall inwards, becoming more and more compressed. This means that the electrons in the interior are continually pushed closer and closer to the nuclei. The ultimate result is that the electrons are forced to tunnel into the nuclei and there interact, forming a gas composed almost entirely of neutrons. Now neutrons, when very close together, repel each other. Consequently, material that could not previously withstand the gravitational force is now given more strength, and can settle into a new, stable equilibrium. Objects attaining this type of equilibrium have sizes measured in miles, and densities a billion times that of water. This means that we can finally identify the stellar remnant from the supernova explosion in the Crab nebula as a *neutron star.*

Several pulsars have now been detected. It has been noted, moreover, that all of them are slowing down: that is, their rotation rate is decreasing. Since the characteristic pulses of radiation are actually powered by the spin of the neutron star, this implies that the pulsar effect must eventually die away. There must therefore be an even larger number of quiescent neutron stars in our neighbourhood. Altogether, neutron stars must be a quite common feature of our Galaxy. Even so, there are limits on the number of stars that can evolve to the neutron star stage. The Chandrasekhar limit occurs because there is a maximum mass that can be sustained by degenerate matter. In a similar way, there is a maximum gravitational pull that can be withstood by the pressure of neutrons. Theoretical estimates of the corresponding limiting mass vary, but a reasonable upper value would be $3M_\odot$. We are therefore faced again with the question — what happens to a dying star that is more massive than this limit?

Black holes

In order to get a rocket away from the Earth it must be given a speed that exceeds a certain minimum value, known as the 'escape velocity'. An escape velocity can similarly be defined for any other object — whether planet, star or galaxy. Its value varies with the size and mass of the object concerned, becoming very high for dense, massive objects. Thus the escape velocity from a neutron star is high. (We can, in fact, turn this statement round. If we could take a stone and drop it onto a neutron star, it would strike the star's surface with this same escape velocity. Such a collision generates heat, the amount generated being dependent on the velocity. For a neutron star, the energy produced by the impact of a small stone would be about the same as that of a small atomic bomb.) We can imagine increasing the mass of a neutron star, or compressing it further, until the escape velocity equals the speed of light. A star of this type would be completely invisible, since no radiation can escape from it. Material can fall into the star, but none can ever escape from it, since it is a basic principle of relativity theory that nothing can travel faster than the speed of light. An object of this sort is therefore called a *black hole*.

Suppose that a supernova explosion leaves behind a fairly massive stellar remnant. It will then violate the requirements for being stable either as a white dwarf, or as a neutron star. All it can do is to collapse inwards, becoming denser and denser, until it reaches a critical point and disappears, forming a black hole. We thus have three ways in which stars can die — as white dwarfs, neutron stars, or black holes. Whereas we can talk about the internal structure and external appearance of the first two, we can obviously not discuss black holes in the same way. This does not mean, however, that black holes are totally beyond the realms of observation. If a black hole accretes material in any quantity, radiation is likely to be produced. The reason is that the material, since it is falling into a very small object, becomes highly compressed. It therefore heats up: indeed, its temperature becomes so high that it emits copious X-rays. These latter can be detected — even if the black hole, itself, cannot — and the existence of a black hole can so be confirmed. At least one likely black hole has now been found — the X-ray object known as Cygnus X-1. The object emitting X-rays is in this case one component of a binary star: it is accreting material from the other component, which appears to be an ordinary star. From observations of the binary motion, tentative masses have been assigned to the two components. The invisible component apparently has a mass well above the neutron star limit of $3M_\odot$ noted above. It therefore fulfils the requirements for a black hole.

CHAPTER 9

Stars and Galaxies

Galactic Evolution

In the preceding chapters we have followed the life histories of individual stars from their birth to their death. But stars are not isolated bodies: they cluster together to form galaxies. We ought, therefore, to end our study by examining briefly the question of galactic evolution. This is equivalent to asking not how an individual star changes with time, but how a group of stars changes as a whole, as successive generations of stars are born and die within it.

We will examine first how our own Galaxy has changed with time. We suppose that originally there was an extremely large cloud of gas, several hundred thousand million times more massive than the Sun, and very diffuse. This cloud differed from those we see at present within our Galaxy, not only in being very much bigger, but also in containing no dust. It may well have consisted only of hydrogen and helium. Under the effect of its own gravitational field the cloud fell inwards — in much the same way as we have described previously for the birth of a star. Its subsequent development also proceeds on similar lines. It contracts, fragments, contracts further, fragments again, and so on. We may, perhaps, identify the first fragments formed with individual galaxies (since galaxies almost invariably occur in groups). We can identify the final fragments, as before, with stars.

The resulting galaxy will contain stars which can be labelled as very extreme examples of Population II. We may suppose that these stars cover a range of masses, and therefore evolve at a variety of rates. The massive ones soon start to burn hydrogen, but, since we have postulated that only hydrogen and helium are present, the carbon cycle cannot operate immediately. Instead, stars of all masses must initially follow the proton-proton chain. However, this makes no significant difference to their overall life-history. The massive stars evolve rapidly, and soon blow up as supernovae. This returns to interstellar space the heavier elements which have been built up in their interiors. We can imagine that the interstellar space still contains clouds of hydrogen and helium left over from the initial burst of star formation. This residual gas is now contaminated by the supernovae ejecta. A second generation of stars forms from the contaminated material. The massive stars of this generation again evolve rapidly, and scatter back still more of the heavier elements into space.

We can thus form a picture of how our supposed galaxy evolves. The amount of gas available for conversion into stars steadily decreases as it is locked up in the low-mass stars which are evolving slowly. On the other hand, the continued explosion of supernovae steadily increases the percentage of elements other than hydrogen and helium in the interstellar medium (and therefore in each new generation of stars). We would expect that an inspection of the stars in such a galaxy should show that they have different proportions of the heavier elements, and that the exact percentage depends on their age. This is precisely the result we have obtained from a study of the stars in our own Galaxy.

Observational Characteristics of Galaxies

What are the characteristics of galaxies which we can use in studying their evolution? Basically, of course, since galactic evolution is an extension of stellar evolution, we would like to study each individual star in a galaxy and determine its individual characteristics along the lines described in earlier chapters. However, apart from a few close galaxies, single stars cannot be distinguished in other stellar systems. (Supernovae form an exception to this rule. They are so bright that they can be seen at very great distances.) What we have to measure, therefore, are characteristic properties associated with entire groups of stars.

The easiest thing to observe, and the most obvious, is the shape of a galaxy. It has been found that a high proportion of all galaxies can be fitted into a very simple classification. This scheme is known as the Hubble classification of galaxies (it was originally introduced by the American astronomer, E. P. Hubble). The galaxies are divided up into three basic groups: ellipticals, spirals and irregulars. Ellipticals are stellar groups which are spheroidal in shape: like a rugger ball, for example. They bear a general resemblance to the globular clusters in our own Galaxy, although they are, of course, much larger. (It must be remembered that we only have a cross-sectional view of these galaxies; we cannot deduce their actual shape immediately.) The spirals resemble our own Galaxy. They all have a central nucleus — rather like an elliptical galaxy, but much flatter — from which streams out long spiral arms. Sometimes the arms, instead of emerging directly from the nucleus, start from the ends of a 'bar' of stars sticking out on either side of the nucleus. This type of galaxy — a barred spiral — is less common than the normal spiral such as our own. Finally, the irregular galaxies, as their name implies, have no special shape, though some of them prove on close inspection to have a certain degree of organization. The sizes of all these galaxies do not seem to depend very greatly on their Hubble type. So one can find large and small elliptical galaxies both with the same apparent shape. The complete Hubble classification is more detailed than this simple division into three groups. The spiral galaxies, for example, are further subdivided according to the size of their nucleus as compared with the extension of their arms.

It is evident from photographs that normal elliptical galaxies have little, if any, dust. Nor do radio and optical observations suggest the presence of great quantities of gas. The nuclei of spiral galaxies generally resemble elliptical galaxies in this respect. On the other hand, the arms of spiral galaxies always contain some dust and gas, and there is even more in irregular galaxies. The 21-centimetre radio observations indicate that irregular galaxies may have nearly 20 per cent of their mass in the form of hydrogen gas. For spiral galaxies the amount is from 10 per cent down to 1 per cent or less. (It depends on their exact Hubble type: the larger figure corresponds to an Sc galaxy and the smaller to an Sa.)

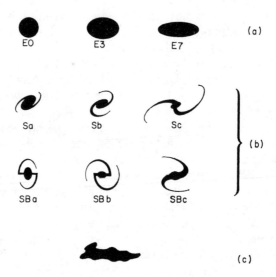

FIG. 55. Types of galaxy: (a) Elliptical, (b) Spiral and Barred Spiral, (c) Irregular (showing their exact Hubble classification).

There is one odd type of galaxy — called an SO — which was not discovered until after the original Hubble classification was created, although it appears to be a fairly common constituent of the Universe. It looks rather like an elliptical galaxy at first glance since it has no spiral arms. But its degree of flattening corresponds to the nucleus of a spiral galaxy, and there are traces of a dust layer all round it.

The extent to which a galaxy is flattened almost certainly

depends on the spin-rate of the galaxy: the faster the spin, the more it is flattened. This can be confirmed from an analysis of the spectra of galaxies. In looking at such spectra we are actually seeing the cumulative light effects of all the stars in the galaxy. Hence we can use them to derive overall characteristics of the galaxy, such as rotation. If we compare the size and shape of the various types of galaxy with their spin-rate, we find that they seem to separate into two distinct groups. On the one hand the spiral and irregular galaxies which rotate fairly rapidly, and on the other the ellipticals with a much lower rotation rate. This difference in rotational energy would suggest that the development and, perhaps, the mode of formation of these two groups are dissimilar.

We have seen that the mass of a star is one of the basic factors influencing its evolution. Similarly, the mass of a galaxy must be of great significance for galactic evolution. Although the determination of galactic masses is even harder than the determination of stellar masses, a certain amount of data have been accumulated during the past few years. It seems that galactic masses increase systematically as we move through the series: Irregular → Sc → Sb → Sa → SO. The elliptical galaxies again form a separate group. They cover a considerable range of masses: the smallest ellipticals are probably the least massive galaxies in the Universe, and the largest are probably the most massive.

In some ways, a more fundamental parameter than the mass of a galaxy is the ratio of its mass to its luminosity. This should depend quite strongly on the most important type of star present. If a galaxy has a large number of very bright stars, this will increase the light emission considerably, but will not have any great effect on the mass. The mass-to-light ratio will therefore be small. If a galaxy has only fainter stars, the light emission will be relatively small, and the mass-to-light ratio will be high. We find, in fact, that the ratio is smallest for the irregulars and the Sc galaxies, and increases towards the SO galaxies. It is generally large in elliptical galaxies, reaching its maximum for the giant ellipticals. We can therefore deduce immediately that the proportion of bright stars is highest in the irregular and Sc galaxies and lowest in the ellipticals, thus following the same distribution as the amount of gas present.

We can confirm this result by studying the spectra of the different types of galaxy. The main spectral features will be determined by the stars which produce the most light, and should therefore vary with the mass-to-light ratio. We do, indeed, find such variations, but they are complicated by yet another factor: the presence of different population types within the galaxy. Our neighbouring Andromeda galaxy, for example, gives one type of spectrum for the nuclear bulge and another for the spiral arms. The difference is what we would expect for Population II stars in the nucleus and Population I stars in the arms, thus showing that the Andromeda galaxy is directly comparable with our own. We may, indeed, deduce from the spectra and mass-to-light ratios available that irregular galaxies contain a large number of young stars, probably comparable with Population I in our Galaxy, whilst elliptical galaxies contain older stars more like Population II.

Peculiar and Normal Galaxies

An appreciable proportion of all galaxies observed do not fit into the simple Hubble classification. The peculiarities may take several different forms. It may be, for example, that two or three galaxies close together distort each other violently: rather as the components of a very close double star are distorted. It may be that an isolated galaxy appears to be undergoing an explosion. In this case, it is usually the centre of the galaxy that is disturbed, and evidence of the 'explosion' is often more readily detectable at radio wavelengths than in the visual. Finally, there are objects that cannot be related to the Hubble classification at all. An obvious example is provided by the quasars. Through the telescope, these objects look just like stars, but they are generally accepted as small, immensely energetic galaxies, probably at very great distances from us.

Clearly, stellar evolution in such peculiar galaxies may follow a different path from that pursued by stars in our own Galaxy. But it is also worth considering whether we really understand our own Galaxy as completely as we should. Observations, particularly radio observations, have shown that our galactic nucleus is the focus of considerable activity. A million solar masses of gas are moving away from the centre at a speed of a hundred miles per second. This is less than we see in 'peculiar' galaxies, but, nevertheless, corresponds to a major 'explosive' event. If, indeed, the gas comes from an explosion in the nucleus, this can only have occurred some ten million years ago — a very short timespan compared with an estimated age for our Galaxy of more than 10,000 million years. We are thus faced with the possibility that all galaxies are to some extent peculiar: in few instances are we clear as to the cause, not even for our own Galaxy. Nor is this all. If we look at our neighbouring galaxies — the members of the Local Group — we can see inexplicable differences in their properties. For example, otherwise similar galaxies may have quite different numbers of star clusters. Until we can explain these differences, we cannot really claim to understand how galaxies evolve.

In this book, we have been tracing how individual stars evolve. Though many uncertainties remain, some kind of consensus on stellar evolution does now exist. The next step is to understand changes in successive generations of stars as galaxies evolve. Here, we still have some way to go.

Further Reading

Most books and articles on stellar evolution expect the reader to have some mathematical background. However, several popular science journals do contain articles written at the level of the present book. Two that frequently contain material on stellar evolution are *Scientific American* and *Sky & Telescope* (both published in the U.S.A.). A regular reading of these journals — especially the latter — will keep the reader up-to-date on new developments.

Some of the older articles on stellar evolution in *Sky & Telescope* have been gathered together in:

T. and L. W. Page (Ed.) *The Evolution of Stars* (Macmillan, N.Y.; 1968).

Readers who can cope with A-level mathematics are recommended to consult two books by R. J. Tayler — *The Stars: their Structure and Evolution* and *The Origin of the Chemical Elements*. Both appeared under the imprint of Wykeham Publications, in 1970 and 1972, respectively. At a slightly more demanding level, V. C. Reddish *The Physics of Stellar Interiors* (Edinburgh University Press; 1974), can be recommended. All these three titles are short. A much more detailed mathematical and physical treatment can be found in such a text as: L. Motz *Astrophysics and Stellar Structure* (Ginn; 1970).

Index

169